A BEAUTIFUL MATH

A BEAUTIFUL MATH

JOHN NASH,

GAME THEORY,

AND THE

MODERN QUEST

FOR A CODE

OF NATURE

TOM SIEGFRIED

JOSEPH HENRY PRESS
Washington, D.C.

Joseph Henry Press • 500 Fifth Street, NW • Washington, DC 20001

The Joseph Henry Press, an imprint of the National Academies Press, was created with the goal of making books on science, technology, and health more widely available to professionals and the public. Joseph Henry was one of the founders of the National Academy of Sciences and a leader in early American science.

Any opinions, findings, conclusions, or recommendations expressed in this volume are those of the author and do not necessarily reflect the views of the National Academy of Sciences or its affiliated institutions.

Library of Congress Cataloging-in-Publication Data

Siegfried, Tom, 1950-
  A beautiful math : John Nash, game theory, and the modern quest for a code
of nature / Tom Siegfried. — 1st ed.
    p. cm.
  Includes bibliographical references and index.
  ISBN 0-309-10192-1 (hardback) — ISBN 0-309-65928-0 (pdfs) 1.
Game theory. I. Title.
  QA269.S574 2006
  519.3—dc22

                          2006012394

Preface

Shortly after 9/11, a Russian scientist named Dmitri Gusev proposed an explanation for the origin of the name Al Qaeda. He suggested that the terrorist organization took its name from Isaac Asimov's famous 1950s science fiction novels known as the Foundation Trilogy. After all, he reasoned, the Arabic word "qaeda" means something like "base" or "foundation." And the first novel in Asimov's trilogy, *Foundation*, apparently was titled "al-Qaida" in an Arabic translation.

In Asimov's books, "Foundation" referred to an organization dedicated to salvaging a decaying galactic empire. The empire was hopeless, destined to crumble into chaos, leaving civilization in ruins for 30,000 years. Foreseeing the inevitability of the empire's demise, one man devised a plan to truncate the coming era of darkness to a mere millennium. His strategy was to establish a "foundation" of scholars who would preserve human knowledge for civilization's eventual rebirth.

At least that's what he told the empire's authorities.

In fact, Asimov's hero, a mathematician named Hari Seldon, created a community of scientists devoted to manipulating the future. Seldon actually formed two foundations—one in a remote but known locale (sort of like Afghanistan), the other in a mystery location referred to only with riddles. Foundation I participated openly in the affairs of the galaxy. Foundation II operated surreptitiously, intervening at key points in history to nudge events along Seldon's chosen path.

Seldon's plan for controlling human affairs was based on a

mathematical system that he invented called psychohistory. It enabled Seldon to predict political, economic, and social trends; foresee the rise and fall of governments; and anticipate the onset of wars and periods of peace.

I don't think Osama bin Laden is Hari Seldon. But it's not so far-fetched to believe that the organizers of the real Al Qaeda perceived Western civilization as an empire in decay. Or that they anointed themselves as society's saviors, hoping to manipulate events in a way that would lead to a new world order more to their liking. So perhaps they adopted some of Hari Seldon's strategies. (Certainly Osama bin Laden's occasional taped messages are eerily similar to Seldon's video appearances from time to time, prepared before his death for delivery decades or even centuries later.)

Of course, any such link to Asimov changes nothing about terrorism. Al Qaeda gains no justification for atrocity from any connection to science fiction. And frankly, the similarities seem rather superficial. Had the terrorists really studied *Foundation*, they would have noticed Asimov's assertion that "violence is the last refuge of the incompetent."

But in fact, Asimov's series did inspire some real-world imitators: not terrorists, but scientists—scientists seeking the secrets of Hari Seldon's psychohistory. If there is a real-life Hari Seldon, it is not Osama bin Laden, but John Forbes Nash.

Nash's life, chronicled so engagingly by Sylvia Nasar in *A Beautiful Mind*, is a story of the struggles of a brilliant but troubled man. Nash's math, for which he won a Nobel Prize, is an entirely different tale, still unfolding, about science's struggle to cope with the complexities of collective human behavior.

At the same time Asimov was publishing his Foundation books, Nash was publishing papers establishing foundational principles for a science called game theory. Game theory is the science of strategy; its formulas tell you what choices to make to get the best deal you can get when interacting with other people. Originally formulated to be applied to economics, game theory has now infiltrated nearly every field of modern science, especially those concerned with human nature and behavior. It has begun to establish links with the physical sciences as well, and ultimately, I suspect, it will forge a merger of all the sciences in the spirit of Asimov's psychohistory. At least that is the prospect that I explore in this book.

Game theory is a rich, profound, and controversial field, and there is much more to it than you could find in any one book. What follows is in no way a textbook on game theory. Nor do I attempt to give any account of its widespread uses in economics, the realm for which it was invented, or the many variants and refinements that have been developed to expand its economic applications. My focus is rather on how various manifestations of game theory built on Nash's foundation are now applied in a vast range of other scientific disciplines, with special attention to those arenas where game theory illuminates human nature and behavior (and where it connects with other fields seeking similar insights). I view these efforts in the context of the ancient quest for a "Code of Nature" describing the "laws" of human behavior, a historical precursor to Asimov's notion of psychohistory.

As with all my books, I try to give any interested reader a flavor of what scientists are doing at the frontiers of knowledge, where there are no guarantees of ultimate success, but where pioneers are probing intriguing possibilities. There are scientists who regard some of this pioneering work as at best misguided and at worst a fruitless waste of time. Consequently, there may be objections from traditionalists who believe that the importance of game theory is overstated or that the prospects for a science of society are overhyped. Well, maybe so. Time will tell. For now, the fact is that game theory has already established itself as an essential tool in the behavioral sciences, where it is widely regarded as a unifying language for investigating human behavior. Game theory's prominence in evolutionary biology builds a natural bridge between the life sciences and the behavioral sciences. And connections have been established between game theory and two of the most prominent pillars of physics: statistical mechanics and quantum theory. Certainly many physicists, neuroscientists, and social scientists from various disciplines are indeed pursuing the dream of a quantitative science of human behavior. Game theory is showing signs of playing an increasingly important role in that endeavor. It's a story of exploration along the shoreline separating the continent of knowledge from an ocean of ignorance, and I think it's a story worth telling.

I owe much gratitude to those who helped make this book possible, particularly the many scientists who have discussed their research with me over the years. Their help is acknowledged by their presence in the pages that follow. Many other friends and colleagues have listened patiently while I've shaped my thoughts on this book during conversations with them. They know who they are, and I appreciate them all. The one person I want to thank by name is my wife, Chris, who really made it possible for me to write this book, because she has a job.

Tom Siegfried
Los Angeles, California

Contents

Introduction

Could not mind, as well as mindless motion, have an
underlying order?

> —Emperor Cleon to Hari Seldon, *Prelude to Foundation*

Isaac Asimov excelled at predicting the future.

In one of his early science fiction stories, he introduced pocket
calculators decades before you could buy them at Radio Shack. In
a later book, he described a digital camera transmitting photos
directly to a computer via WiFi.[1] He just forgot to mention that
you could also use the same device to make phone calls. And in his
most celebrated work, a series of 1950s science fiction novels
known as the Foundation Trilogy, Asimov foresaw a new kind of
science called psychohistory, capable itself of forecasting political,
economic, and social events. Psychohistory, as Asimov envisioned
it, was "the science of human behavior reduced to mathematical
equations."[2]

Real-life psychohistory does not yet exist—not now, not really, and not for a long time. But there are many research enterprises under way in the world today that share the goal of better
understanding human behavior in order to foresee the future. At
the foundation of these enterprises are mathematical methods
closely resembling Asimov's psychohistory. And in the midst of it
all is the work of a mathematician named John Forbes Nash.

Brilliant but odd, intellectually sophisticated but socially awkward, Nash dazzled the world of mathematics in the 1950s with astounding and original results in several arenas. He rattled the routines at Princeton University and the Rand Corporation in California with both his mental magnificence and his disruptive behavior. By now, the subsequent tragic aspects of Nash's life story are familiar to millions of people, thanks to the Oscar-winning movie starring Russell Crowe, and Sylvia Nasar's *A Beautiful Mind*, the acclaimed book on which the movie was based. Yet while book and movie probed the conflicting complexities in Nash the man, neither delved deeply into Nash's math. So for most people today, his accomplishments remain obscure. Within the world of science, though, Nash's math now touches more disciplines than Newton's or Einstein's. What Newton's and Einstein's math did for the physical universe, Nash's math may now be accomplishing for the biological and social universe.

Indeed, had mental illness not intervened, Nash's name might today be commonly uttered in the same breath with those scientific giants of the past. As it is, he made important contributions to a few mathematical specialties. But he achieved his greatest fame in economics, the field in which he shared the 1994 Nobel Prize with John Harsanyi and Reinhard Selten for their seminal work on the theory of games—the math that analyzes how people make choices in contests of strategy.

Game theory originated in efforts to understand parlor games like poker and chess, and was first fully formulated as a mathematical tool for describing economic behavior. But in principle, game theory encompasses any situation involving strategic interaction—from playing tennis to waging war. Game theory provides the mathematical means of computing the payoffs to be expected from various possible choices of strategies. So game theory's math specifies the formulas for making sound decisions in any competitive arena. As such, it is "a tool for investigating the world," as the economist Herbert Gintis points out. But it is much more than a mere tool. "Game theory is about how people cooperate as much as how they compete," Gintis writes. "Game theory is about the emergence, transformation, diffusion and stabilization of forms of behavior."[3]

Nash did not invent game theory, but he expanded its scope and provided it with more powerful tools for tackling real-world problems. At first, though, the depth of his accomplishment was little appreciated. When his revolutionary papers appeared, in the early 1950s, game theory briefly became popular among Cold War analysts who saw similarities between international aggression and maximizing profits. But within economics, game theory remained mainly a curiosity. "It didn't take off," the economist Samuel Bowles told me. "Like a lot of good ideas in economics, it just fell by the wayside."[4]

In the 1970s, though, evolutionary biologists adopted game theory to study the competition for survival among animals and plants. And in the 1980s, economists finally began to use game theory in various ways, finding it especially helpful in designing actual experiments to test economic theory. By the late 1980s game theory had re-emerged in economics in a big way, leading to Nash's 1994 Nobel.

Even before then, game theory had already migrated into the curricula of many scientific disciplines. You could find it taught in departments not only of mathematics and economics and biology but also political science, psychology, and sociology. By the opening years of the 21st century, game theory's uses had spread even wider, to fields ranging from anthropology to neurobiology.

Today, economists continue to use game theory to analyze how people make choices about money. Biologists apply it to scenarios explaining the survival of the fittest or the origin of altruism. Anthropologists play games with people from primitive cultures to reveal the diversity of human nature. And neuroscientists have joined the fun, peering inside the brains of game-playing people to discover how their strategies reflect different motives and emotions. In fact, a whole new field of study, called neuroeconomics, has taken shape, mixing game theory's methods with brain-scanning technology to detect and measure neural activity corresponding to human judgments and behavior. "We're quantifying human experience," says neuroscientist Read Montague, "in the same way we quantify airflow over the wings of a Boeing 777."[5]

In short, Nash's math—with the rest of modern game theory built around it—is now the weapon of choice in the scientist's arsenal on a wide range of research frontiers related to human behavior. In fact, Herbert Gintis contends, game theory has become "a universal language for the unification of the behavioral sciences."[6]

I think it might go even farther than that. Game theory may become the language not just of the behavioral sciences, but of all the sciences.

As science stands today, that claim is rather bold. It might even be wrong. But game theory already has conquered the social sciences and invaded biology. And it is now, in the works of a few pioneering scientists, forming a powerful alliance with physics. Physicists, of course, have always sought a unity in the ultimate description of nature, and game theory may have the potential to be a great unifier.

That realization hit me in early 2004, when I read a paper by physicist-mathematician David Wolpert, who works at NASA's Ames Research Center in California. Wolpert's paper disclosed a deep connection between the math of game theory and statistical mechanics, one of the most powerful all-purpose tools used by physicists for describing the complexities of the world.

Physicists have used statistical mechanics for more than a century to describe such things as gases, chemical reactions, and the properties of magnetic materials—essentially to quantify the behavior of matter in all sorts of circumstances. It's a way to describe the big picture when lacking data about the details. You can't track every one of the trillion trillion molecules of air zipping around in a room, for instance, but statistical mechanics can tell you how an air conditioner will affect the overall temperature.

It's no coincidence that statistical mechanics (which encompasses the kinetic theory of gases) is the math that inspired Asimov's heroic mathematician, Hari Seldon, to invent psychohistory. As Janov Pelorat, a character in the later novels of the Foundation series, explained:

> Hari Seldon devised psychohistory by modeling it upon the kinetic theory of gases. Each atom or molecule in a gas moves randomly so that we can't know the position or velocity of any one of them. Nevertheless, using statistics, we can work out the rules governing their overall behavior with great precision. In the same way, Seldon intended to work out the overall behavior of human societies even though the solutions would not apply to the behavior of individual human beings.[7]

In other words, put enough people together and the laws of human interaction will produce predictable patterns—just as the interactions and motion of molecules determine the temperature and pressure of a gas. And describing people as though they were molecules is just what many physicists are doing today—in effect, they're taking the temperature of society.

One of the best ways to take that temperature, it turns out, is to view society in terms of networks. In much the same way that "temperature" captures an essential property of a jumble of gas molecules, network math quantifies how "connected" the members of a social group are. Today's new network math applies statistical mechanics to all sorts of social phenomena, from fashion trends and voting behavior to the growth of terrorist cells. So just as Asimov envisioned, statistical physics has been enlisted to describe human society in a mathematically precise way.

For the most part, this merger of network math and statistical mechanics has been exploring human behavior without recourse to the modern views of game theory built on Nash's math. After all, Nash's original formulation had its limits; what works on paper does not always play out the way his math predicts in real-world games. But the latest research has begun to show ways that game theory can help make sense out of the intricate pattern of links in complicated networks. The game theory approach may be able to induce the world of complex networks to more readily surrender its secrets.

Wolpert's insight suggests that game theory itself can be elevated to a new level by exploiting its link to statistical mechanics. His work shows that the math of game theory can be recast in equations that mimic those used by statistical physicists to describe all sorts of physical systems. In other words, at some deep level statistical mechanics and game theory are, in a sense, two versions of the same underlying idea. And that may end up making game theory an especially sensitive social thermometer.

This new realization—that game theory and statistical mechanics share a deep mathematical unity—enhances game theory's status as the preferred tool for merging the life sciences and physical sciences into a unified description of nature. After all, there's a reason why game theory has been embraced by so many disciplines. Game theory could someday become the glue that holds all of science's puzzle pieces together.

Some people (particularly many physicists) will scoff at this contention. But pause to consider how much sense it makes. Nature encompasses so many complex networks for a reason: complexity evolves. "Intelligent" design produces simple, predictable systems that are easy to understand. The complex systems that baffle science—like bodies, brains, and societies—arise not from any plan, but from interactions among agents like cells or people, all (more or less) out for themselves. And such competitive interaction is precisely what game theory is all about.

So it should not be surprising that game theory has been so useful in evolutionary biology. Game theory is about competition, and evolution is the ultimate never-ending Olympic event. And if evolution followed game theory's rules in generating complicated life, it no doubt also observed the same rules in developing the human brain. So it's perfectly natural that game theory has become popular today in efforts to understand how the brain works, as brain scientists explore the neural physiology behind economic choices.

In turn, the brain underlies all the rest of human behavior—personal and interpersonal, social and political, as well as economic. All that behavior directs the evolution of all those networks of personal, social, political, or economic activity. Just as the complexities of life arose through eons of survival of the fittest, human culture evolves as societies or governments rise and fall; economies evolve as companies are founded and go bankrupt; even the World Wide Web evolves as pages are added and links expire. So Nash's math does seem capable of catalyzing a merger of methods for understanding individual behavior, biology, and society.

What about chemistry and physics? At first glance there doesn't seem to be any struggle for survival among the molecules engaged in chemical reactions. But in a way there is, and the connections between game theory and statistical mechanics promise to reveal ways in which game theory still applies. Reacting molecules, for instance, always seek a stable condition, in which their energy is at a minimum. The "desire" for minimum energy in molecules is not so different from the "desire" for maximum fitness in organisms. They can be treated mathematically in a similar way.

True, there's much more to physics than statistical mechanics. At first glance, game theory does not seem to touch some of the grander arenas of physical science, such as astrophysics and cosmology, or the subatomic realm ruled by quantum physics. But guess what? In the past few years physicists and mathematicians have developed quantum versions of game theory. So far, quantum theory seems to be enriching game theory, but that enrichment just might turn out to be mutual.[8]

Furthermore, Wolpert forges the link between statistical mechanics and game theory with help from the mathematical theory of information. As I wrote in my book *The Bit and the Pendulum* (Wiley, 2000), modern science has become enamored of information theory, using both its math and its metaphor to describe all sorts of science, from the contents of black holes to the computational activity in the brain. Quantum physics itself has been illuminated over the past decade by new insights emerging from quantum information theory. And some theorists have pursued the notion that information ideas hold the key to unifying quantum physics with gravity, perhaps paving the way to the ultimate "theory of everything." It's possible, Wolpert speculates, that game theory is the ingredient that could enhance the prospects for success in finding such a theory.

In any case, it's already clear that Nash's math shows an unexpectedly powerful way of mirroring the regularities of the real world that make all science possible. As I described in my book *Strange Matters* (Joseph Henry, 2002), there is something strange about the human brain's ability to produce math that captures deep and true aspects of reality, enabling scientists to predict the existence of exotic things like antimatter and black holes before any observer finds them. Part of the solution to that mystery, I suggested, is the fact that the brain evolved in the physical world, its development constrained by the laws of physics as much as by the laws of biology. I failed then to realize that game theory offers a tool for describing how the laws of physics and biology are related.

It's clear now that game theory's math describes the capability of the universe to produce brains that can invent math. And math in turn, as Asimov envisioned, can be used to describe the behavior guided by those brains—including the social collective behavior that creates civilization, culture, economics, and politics.

While seeking the secrets of that math, we can along the way watch people play games as neuroscientists monitor the activity in their brains; we can follow anthropologists to the jungle where they test the game-playing strategies of different cultures; we can track the efforts of physicists to devise equations that capture the essence of human behavior. And just maybe we'll see how Nash's math can broker the merger of economics and psychology, anthropology and sociology, with biology and physics—producing a grand synthesis of the sciences of life in general, human behavior in particular, and maybe even, someday, the entire physical world. In the process, we should at least begin to appreciate the scope of a burgeoning research field, merging the insights of Nash's 1950s math with 21st-century neuroscience and 19th-century physics to pursue the realization of Asimov's 1950s science fiction dream.

It would be wrong, though, to suggest that Asimov was the first to articulate that dream. In a very real sense, psychohistory was the reincarnation of the old Roman notion of a "Code of Nature" (fitting, since Asimov's Foundation series was modeled on the Roman Empire's decline and fall). As interpreted much later, that code supposedly captured the essence of human nature, providing a sort of rule book for behavior. It was not a rule book in the sense of prescribing behavior, but rather a book revealing how humans naturally behave. With the arrival of the Age of Reason in the 18th century, philosophers and the forerunners of social scientists sought in earnest to discover that code of codes—the key to understanding the natural order of human interaction. One of the earliest and most influential of those efforts was the economic system described in *The Wealth of Nations* by Adam Smith.

1

Smith's Hand

Searching for the Code of Nature

> If in the seventeenth century natural philosophers
> borrowed notions of law in human affairs and ap-
> plied them to the study of physical nature, in the
> eighteenth century it was the turn of the laws of
> physical nature to suggest ways forward for knowl-
> edge about human life.
>
> —Roger Smith, *The Norton History of the Human Sciences*

Colin Camerer was a child prodigy, one of those kids who skipped
several grades of school and enrolled in a special program for the
gifted. By age 5, he was reading *Time* magazine (even though no
one had taught him to read), and at 14 he entered Johns Hopkins
University. He graduated in three years, then went to the University of Chicago to earn an M.B.A. and, for good measure, a Ph.D.
He joined the faculty at Northwestern University's graduate school
of management by the age of 22.

Today, he's a full-fledged adult on the faculty at Caltech, where
he likes to play games. Or more accurately, he likes to analyze the
behavior of other people during various game-playing experiments. Camerer is one of the nation's premier behavioral game
theorists. He studies how game theory reveals the realities of human economic behavior, how people in real life depart from the
purely rational choices assumed by traditional economic theory.

Though unquestionably brilliant, Camerer communicates as conversationally as a cab driver. Even in his prodigy days, he was a wrestler and a golfer, so he has a broader view of the world than some of the intellectually exalted scholars who live their lives on such a higher mental level. And he has a broader view of economics than you'll find in the old-school textbooks. But in a sense, Camerer's views on economic behavior are not so revolutionary. In fact, in some ways they were anticipated by the father of traditional economics, Adam Smith.

Smith's "invisible hand" is probably the most famous metaphor in all of economics, and his equally famous book, *Wealth of Nations*, remains revered by today's advocates of free-market economies more than two centuries after its publication. But Smith was not a one-dimensional thinker, and he understood a lot more about human behavior than many of his present-day disciples do. His insights foreshadowed much in current attempts to decipher the code of human conduct, in economics and other social arenas. He was not a game theorist, but his theories illuminate the links between games, economics, biology, physics, and society—which is what the book you're reading now is all about. The way I see it, Adam Smith was the premier player in the origins of this story, as he inspired belief in the merit of melding the Newtonian physics of the material world with the science of human behavior.

THE ECONOMICS OF INVISIBILITY

Adam Smith had a lot in common with Isaac Newton. Both were lifelong bachelors. Both became professors at the university they had attended (and both had reputations for being absentminded professors as well). Both were born after their fathers had died. And both became fathers themselves of a new scientific discipline. Newton built the foundation of physics; Smith authored the bible of economics.

Both men literally rewrote the book of their science, transforming the somewhat inchoate insights of their predecessors into treatises that guided modern thought. Just as modern physics de-

scended from Newton's codification of what was then known as
natural philosophy, modern economics is the offspring of Adam
Smith's treatise on political economy. And though their major
works were separated by nearly a century, the philosophies they
articulated merged to forge a new worldview coloring virtually
every aspect of European culture in the centuries that followed.

While Newton established the notion of natural law in the
physical world, Smith tried to do the same in the social world of
economic intercourse. Newton's unexplained law of gravity
reached across space to guide the motion of planets; Smith's "invisible hand" guided individual laborers and businessmen to produce the wealth of nations. Together, Newton's and Smith's works
inspired great thinkers to believe that all aspects of the world—
physical and social—could be understood, and explained, by science. When Smith's *Wealth of Nations* was published in 1776, the
Age of Reason reached its pinnacle.

Nowadays, of course, physics has moved beyond Newton, and
most economists would say that their science has moved far beyond Adam Smith. But Smith's imprint on modern culture persists,
and his impact on economic science remains substantial. If you
look closely, you can even find echoes of Smith's ideas in various
aspects of game theory.

For one thing, Smith ingrained the idea that pursuing
self-interest drives economic prosperity. And it is pursuit of self-
interest that game theory, at its most basic level, attempts to quantify. At a deeper level, Smith sought a system that captured the
essence of human nature and behavior, a motivation shared by
many modern game theorists. Game theory tries to delimit what
rational behavior is; Smith helped deposit the idea in the modern
mind that minds operate in a rational way.

It was one thing for Newton to assert that rational laws governed the motions of the planets or falling apples. It was much
more ambitious for Smith to ascribe similar orderliness to the social behavior of humans engaging in economic activity. As Jacob
Bronowski and Bruce Mazlish observed in a now old, but still
insightful, book on Western thought, Smith took a bit of an intel-

lectual leap to make his system fly. "In order to discover such a science as economics," they wrote, "Smith had to posit a faith in the orderly structure of nature, underlying appearances and accessible to man's reason."[1]

Viewed in these terms, Smith's book was an important thread in a fabric of thought seeking a Code of Nature, a system of rules that explained human behavior (economic and otherwise) in much the same way that Newton had explained the cosmos. First philosophers, and then later sociologists and psychologists, tried to articulate a science of human behavior based on principles "underlying appearances" but "accessible to man's reason." Smith's efforts reflected the influence of his friend and fellow Scotsman David Hume, the historian-philosopher who regarded a "science of man" as the ultimate goal of the scientific enterprise. "There is no question of importance, whose decision is not comprised in the science of man," Hume wrote, "and there is none, which can be decided with any certainty, before we become acquainted with that science."[2] In the attempt "to explain the principles of human nature, we in effect propose a compleat system of the sciences."

Today, game theory's ubiquitous role in the human sciences suggests that its ambitions are woven from that same fabric. Game theory may, someday, turn out to be the foundation of a new and improved 21st-century version of the Code of Nature, fulfilling the dreams of Hume, Smith, and many others in centuries past.

That claim is enhanced, I think, with the realization that threads of Smith's thought are entangled not only in physical and social science, but biological science as well. Smith's ideas exerted a profound influence on Charles Darwin. Principles describing competition in the economic world, Darwin realized, made equal sense when applied to the battle for survival in the biological arena. And the benefits of the division of labor among workers that Smith extolled meshed nicely with the appearance of new species in nature. So it is surely no accident that, today, applying economic game theory to the study of evolution is a major intellectual industry.

LOGIC AND MORALS

All in all, Smith's economics provides a critical backdrop for understanding the economic world that game theory conquered in the 20th century. His influence on today's world stemmed from a life spent gathering unusual insights into his own world.

Born in Scotland in 1723, Smith was a sickly weakling as a child (today we'd probably call him athletically challenged). At the age of 3, he was kidnapped from his uncle's front porch by some gypsylike vagrants known as tinkers. Apparently the uncle rescued the toddler shortly thereafter. Growing up, Adam was a bright kid, earning a reputation as a bookworm with a spectacular memory. At 14 he entered the University of Glasgow (in those days, that was not unusually young). At 17 he went to Oxford, at first with the intention of entering the clergy. But after seven years there he returned to Scotland in search of a different kind of life. His interests destined him to the academic world, as he had no acumen for business and, as one biographer noted, "a strong preference for the life of learning and literature over the professional or political life."[3]

After a time, Smith got the job that fit his interests and talents—professor of logic at the University of Glasgow. Soon he was also appointed to a professorship in "moral philosophy," providing a fitting combination of duties for someone planning to forge a rational understanding of human behavior. It was, in fact, moral philosophy that Smith seized on for his first significant treatise. And in it he outlined a very different view of life and government than what he is generally known for today. His book on morals won him the confidence of Charles Townsend, who employed Smith to tutor his stepson, the young Duke of Buccleuch. Smith left Glasgow for London in 1764 to assume his tutorial task. He and the duke traveled much during this tutorship, spending a lot of time in France, where Smith familiarized himself with the new economic ideas of a group known as the physiocrats.

Smith was especially taken with one François Quesnay, a fascinating character who deserves to be much better known than he is.

Orphaned from working-class parents (some sources say farmers) at 13, Quesnay taught himself to read using a medical book, and so decided he might as well become a doctor. He established himself as a physician and became an early advocate for surgery as an important part of medical practice—not such a popular position among doctors of his day. Quesnay played a part, though, in getting the King of France to separate surgeons from barbers, surely a benefit for both professions. Quesnay's even stronger influence with King Louis XV was later secured when he attained an appointment as personal physician to Madame de Pampadour, the king's mistress.

Quesnay must have possessed an unusually fine mind; he impressed his patients dramatically, generating the word of mouth that led to such connections in high places. Once established among the aristocracy, Quesnay's brilliance attracted the other leading intellects of his age, so much so that he was invited to write articles on agriculture for the famous French *Encyclopédie*. Somewhere along the way his agricultural interest led to an interest in economic theory, and Quesnay founded the new school of economists whose practitioners came to be called the physiocrats, out of their affinity for the methods of physics.

In those days, conventional wisdom conceived of a nation's economic strength in terms of trade; favorable trade balances, therefore, supposedly brought wealth to a nation. But Quesnay argued that the true source of wealth was agriculture—the productivity of the land. He further argued that governments imposed a human-designed impairment to the "natural order" of economic and social interaction. A "laissez-faire" or "hands-off" policy should be preferred, he believed, to allow the natural flow to occur.

Encountering Quesnay while in Paris, Smith was also entranced and began to merge the physiocratic philosophy with his own. Upon his return to England in 1766, Smith embarked on the decade-long task of compiling his insights into human nature and the production of prosperity, ending with the famous tome titled *An Inquiry into the Nature and Causes of the Wealth of Nations,* mercifully shortened in casual usage to simply *Wealth of Nations.*

THE INVISIBLE HAND

Smith's views differed from Quesnay's in one major respect: The source of wealth, Smith argued, was not the land, but labor. "The annual labour of every nation is the fund which originally supplies it with all the necessaries and conveniences of life," Smith declared in his book's Introduction. And the production of wealth was enhanced by dividing the labor into subtasks that could be performed more efficiently using specialized skills. "The greatest improvement in the productive powers of labour, and the greater part of the skill, dexterity, and judgment with which it is any where directed, or applied, seem to have been the effects of the division of labour," Smith pronounced at the beginning of Chapter 1.[4]

Modern caricatures of *Wealth of Nations* do not do it justice. It is usually summed up with a reference to the "invisible hand" that makes capitalism work just fine as long as government doesn't get involved. There is no need for any planning or external economic controls—if everyone simply pursues profits without restraint, the system as a whole will be most efficient at distributing goods and services. With his "invisible hand" analogy, Smith seems to assert that pure selfishness serves the world well: "It is not from the benevolence of the butcher, the brewer, or the baker, that we can expect our dinner, but from their regard to their own interest," Smith wrote. "By directing that industry in such a manner as its produce may be of the greatest value, he intends only his own gain, and he is in this, as in many other cases, led by an invisible hand to promote an end which was no part of his intention."[5]

In fact, Smith's ideas about a free-market economy were subtle and sophisticated, much more thoughtful than the knee-jerk free-market-to-the-max mantra that people promote, invoking his name, today. (Among other things, he noted that the invisible hand worked effectively only if the people doing business weren't crooks cooking the books.) He did believe that government interference in business—either to assist or restrain—subverted the benefits of natural and free enterprise. By eliminating both preferences (or "encouragements") and restraints, "the obvious and simple system of natural liberty establishes itself of its own accord." But even then he restricted his concern to "extraordinary encouragements" or "extraordinary restraints." And he cited three specific roles that government ought to fulfill: defending the country from invasion, enforcing the laws so as to protect individuals from injustice, and providing for the public works and institutions that private individuals would not find profitable (like protecting New Orleans from hurricanes).

Modern economists have noted that Smith's devotion to the invisible hand was expressed in rather qualified language. "There can be little doubt that Smith's faith in the power of an invisible hand has been exaggerated by modern commentators," Princeton economist Alan Krueger wrote in an introduction to a recent reprinting of *Wealth of Nations*.[6] Besides, Krueger added, "most of postwar economics can be thought of as an effort to determine theoretically and empirically when, and under what conditions, Adam Smith's invisible hand turns out to be all thumbs."[7]

All this is not to say that Smith's support for free enterprise is entirely a misreading. (Nor am I saying that free enterprise is exactly a bad idea.) But as economists who followed Smith often observed, his invisible hand *does not always* guarantee efficient markets or fairness. A critique by Thomas Edward Cliffe Leslie, an economic historian in Belfast, about a century after *Wealth of Nations* appeared, noted that Smith wrote in a preindustrial age. However deep his insights into the world he lived in, Smith was nevertheless incapable of escaping his own time.

Some of Smith's followers, Cliffe Leslie wrote, considered *Wealth of Nations* not just an "inquiry," as Smith's full title suggested, but "a final answer to the inquiry—a body of necessary and universal truth, founded on invariable laws of nature, and deduced from the constitution of the human mind." Cliffe Leslie demurred: "I venture to maintain, to the contrary, that political economy is not a body of natural laws in the true sense, but an assemblage of speculations and doctrines . . . colored even by the history and character of its chief writers."[8]

Cliff Leslie's account, published in 1870, dismissed the idea—

promoted by many of Smith's disciples—that Smith had revealed "a natural order of things," an "offshoot of the ancient fiction of a Code of Nature."

This idea of a "code of natural law" had been around since Roman times, with possible Greek antecedents. The Roman legal system recognized not only Roman civil law (*Jus Civile*), the specific legal codes of the Romans, but a more general law (*Jus Gentium*), consisting of laws arising "by natural reason" that are "common to all mankind," as described by Gaius, a Roman jurist of the second century A.D.

Apparently some Roman legal philosophers regarded *Jus Gentium* as the offspring of a forgotten "natural law" (*Jus Naturale*) or "Code of Nature"—an assumed primordial "government-free" legal code shared by all nations and peoples. Human political institutions, in this view, disturb "a beneficial and harmonious natural order of things." So as near as I can tell, "Code of Nature" is what people commonly refer to today as the law of the jungle.[9] (Perhaps the FOX network will develop it as the next new reality-TV series.) "The belief gradually prevailed among the Roman lawyers that the old *Jus Gentium* was in fact the lost code of Nature," English legal scholar Henry Maine wrote in an 1861 treatise titled *Ancient Law*. "Framing . . . jurisprudence on the principles of the *Jus Gentium* was gradually restoring a type from which law had only departed to deteriorate."[10]

In any event, as Cliffe Leslie recounted, the "Code of Nature" idea was, in Smith's day, one of two approaches to grasping "the fundamental laws of human society." The Code of Nature method sought to reason out the laws of society by deducing the natural order of things from innate features of the human mind. The other approach "induced" societal laws by examining history and features of real life to find out how things actually are, rather than some idealized notion of how human nature should be.

In fact, Smith's work did express sentiments favorable to the Code of Nature view; his statement that eliminating governmental preferences and restraints allows "the obvious and simple system of natural liberty" to establish itself clearly resonates with the concept of such a code. And Dugald Stewart, in a biographical memoir of Smith, asserted that Smith's "speculations" attempted "to illustrate the provisions made by Nature in the principles of the human mind" for gradual augmentation of national wealth and "to demonstrate that the most effectual means of advancing a people to greatness is to maintain that order of things which Nature has pointed out."[11]

Cliffe Leslie maintained, on the other hand, that Smith actually pursued both methods—some deductive reasoning, to be sure, but also thorough observations of actual economic conditions of his day. While Smith might have believed himself to be articulating the natural laws of human economic behavior—a Code of Nature—in fact he just developed another human-invented system colored by culture and history, Cliffe Leslie declared.

"What he did not see was, that his own system . . . was the product of a particular history; that what he regarded as the System of Nature was a descendant of the System of Nature as conceived by the ancients, in a form fashioned by the ideas and circumstances of his own time," Cliffe Leslie wrote of Smith. "Had he lived even two generations later, his general theory of the organization of the economic world . . . would have been cast in a very different mould."[12]

If Smith's Code of Nature was tainted by his times, it was nevertheless in tune with many similar efforts by others, before him and after. Various versions of such an idea—the existence of a "natural order" of human behavior and interaction—influenced all manner of philosophers and scientists and political revolutionaries seeking to understand society, everybody from the monarchist philosopher Thomas Hobbes to the science-fan and journalist Karl Marx. Smith's two great works, on moral philosophy and the laws of wealth, were really part of one grander intellectual enterprise that ultimately produced both economics and the "human sciences" of sociology and psychology. As science historian Roger Smith has pointed out, the 18th century—Adam Smith's century—was a time of profound intellectual mergers, with the physical sciences and the social world, economic interaction and human nature, all mutually inspiring new views of understanding and explaining life, the universe, and everything.

"In the eighteenth century," Roger Smith writes, "pleasure and confidence in the design of the created physical world played an important part in the search for the design of the human world." Just as Newton discovered the "natural order" of the physical universe, thinkers who followed pursued the principles behind the "natural order" of society. In fact, the forerunner of economics, political economy, emerged in the last half of the 18th century as "the study of the link between the natural order and material prosperity," investigating "the laws, physical and social, that underlie wealth."[13]

And of course, precisely the same sort of merger fever afflicts scientists today. The mix of math and physics with biology, sociology, and economics is (to use economic terminology) a growth industry, and game theory is becoming the catalyst accelerating the trend.

RATIONAL ISN'T NATURAL

There's an additional subtle point about all this that's essential for understanding the relationship between Smith's ideas and modern notions of human nature and game theory. The cartoon view of Smith's story is that human nature is selfish, and that economic behavior is rooted in that "truth." And game theory seems to incorporate that belief. In its original form, game theory math describes "rational" behavior in a way that essentially synonymizes "rational" with "selfish." But as it is interpreted today, game theory does not actually *assume* that people always behave selfishly—or rationally. Game theory tells you what will happen if people *do* behave selfishly and rationally.

Besides, Adam Smith did *not* believe that humans are universally selfish (and he was right, as game theory experiments have recently rediscovered). In fact, Smith glimpsed many findings of today's experimental economic science. Modern commentators often don't realize that, though, because they neglect to consider that *Wealth of Nations* was not Smith's only book.

When writing *Wealth of Nations*, Smith assumed (as do all authors) that its readers would have also read his first book: the *Theory of Moral Sentiments*, published in 1759. So he did not think it necessary to revisit the much different picture of human nature he had previously presented. Read together, Smith's two books show that he had a kinder and gentler view of human nature than today's economics textbooks indicate.

This point was made to me by Colin Camerer, whose research is at the forefront of understanding the connections between game theory and human behavior. Camerer's specialty, "behavioral game theory," is a subdivision of the field generally described as "behavioral economics." By the 1980s, when game theory began to infiltrate the economics mainstream, various economists had become disenchanted with the old notion, descended with mutations from Adam Smith, that humans were merely rational actors pursuing profits. Some even hit on the bright idea of testing economic theory by doing experiments, with actual people (and sometimes even real money). Not surprisingly, experiments showed that people often act "irrationally"—that is, their choices do not always maximize their profits. Pursuing such experiments led to some Nobel prizes and some new insights into the mathematics underlying economic activity.

Game theory played a central role in those developments, as it quantified the profit maximization, or "utility," that people in experiments were supposed to be pursuing. In a complicated experiment, it's not always obvious what the utility-maximizing strategy really is. Game theory can tell you. In any event, Camerer finds it fascinating that game theory shows, in so many ways, that humans defy traditional economic ideas. But those experimental results, he told me, don't really defy Adam Smith.

During one of our conversations, at a coffee shop on the Caltech campus, Camerer stressed that Smith never contended that all people are inherently selfish, out for themselves with no concern for anyone else. Smith merely pointed out that *even if* people operated totally out of selfishness, the economic system could still function efficiently for the good of all. "The idea was, if people want to make a lot of dough, the way to do it is by giving you what you want, and they don't care about you per se. And that doesn't logically imply that people don't care about others; it just means that even if they didn't, you could have an effective capitalist economy and produce what people most want," Camerer said. "I think Adam Smith has been kind of misread. People say, 'Gee, Adam Smith proved that people don't care about each other.' What he conjectured, and later was proved mathematically, was that even if people didn't care about each other, markets could do a pretty good job of producing the right goods. But logically that doesn't imply that people don't care."[14] So human nature is not necessarily as adamantly self-serving as some people would like to believe. Some people are selfish, of course, but others are not.

In fact, in Smith's treatise on moral sentiments, he identified sympathy as one of the most important of human feelings. And he described the conflict between the person's "impartial spectator"— a sort of long-term planner or "conscience"—and the passions, including hunger, fear, anger, and other drives and emotions. The brain's impartial spectator weighs the costs and consequences of actions, encouraging rational choices that should control the reactions of the passions. While economists have traditionally assumed that people make rational economic choices, Smith knew that in real life the passions often prevailed. "Smith recognized . . . that the impartial spectator could be led astray or rendered impotent by sufficiently intense passions," Camerer and two colleagues wrote in a 2005 paper.[15]

Nevertheless, the notion of self-interest and utility was dramatized by Smith in such a way that it formed a central core of subsequent economic philosophy. And not only economics was shaped by Smith's ideas. His books also contributed in a significant way to the birth of modern biology.

ORIGIN OF DARWINISM

I'm not sure whether Charles Darwin ever read *Wealth of Nations*. But he certainly read accounts of it, including Dugald Stewart's eulogy-biography of Adam Smith. And Darwin was familiar with Smith's *Moral Sentiments*, citing its "striking" first chapter in a passage in *Descent of Man*. And while Darwin's *Origin of Species* does not mention Smith, its notion of natural selection and survival of the fittest appears to be intellectually descended from Smith's ideas of economic competition.

Smith's influence on Darwin was pointed out more than two decades ago by the science historian Silvan Schweber. I first encountered the connection, though, in the late Stephen Jay Gould's massive tome on evolutionary biology. Gould examined Darwin's writings inside and out and traced all sorts of historical, philosophical, scientific, and literary influences on the origin of Darwin's views on origins. Among the most intriguing of those influences was the work of William Paley, the theologian often cited today by supporters of creationism and intelligent (sic) design.

Paley is most famous for his watchmaker analogy. If you find a watch on the ground, Paley wrote in 1802, you can see that it's nothing like a rock. The watch's parts are clearly "put together for a purpose," adjusted to produce "motion so regulated as to point out the hour of the day." The inevitable inference, Paley concluded, was "the watch must have a maker . . . who comprehended its construction, and designed its use." Paley's point was that the biological world was so full of orderly complexity, exquisite adaptation to the needs of efficient living, that it must have been the product of an exquisite design, and hence, a designer. To arrive at his own evolutionary theory, Darwin required an alternative logic to explain the efficiency of life. Adam Smith, Gould concluded, supplied that logic.

"In fact, I would advance the even stronger claim that the theory of natural selection is, in essence, Adam Smith's economics transferred to nature," Gould wrote. "Individual organisms engaged in the 'struggle for existence' act as the analog of firms in competition. Reproductive success becomes the analog of profit."[16] In other words, as Smith argued, there is no need to design an efficient economy (and in fact, a designer would be a bad idea). The economy designs itself quite well if left alone, so that the individuals within that economy are free to pursue their self-interest. Darwin saw a similar picture in biology: Organisms pursuing their own interest (survival and reproduction) can create, over time, complexities of life that mirror the complexities of an economy. In one passage, Darwin refers specifically to the concept of "division of labor," a favorite topic of Smith's. In his famous example of the pin factory, Smith described how specialization breeds efficiency. It seemed to Darwin quite analogous to the origin of new species in nature.

"No naturalist doubts the advantage of what has been called the 'physiological division of labour'; hence we may believe that it would be advantageous to a plant to produce stamens alone in one flower or on one whole plant, and pistils alone in another flower or on another plant," Darwin wrote in *Origin of Species*. Similar advantages of such specialization, he noted, apply to diversity among organisms.

"We may, I think, assume that the modified descendants of any one species will succeed by so much the better as they become more diversified in structure, and are thus enabled to encroach on places occupied by other beings," Darwin commented. "So in the general economy of any land, the more widely and perfectly the animals and plants are diversified for different habits of life, so will a greater number of individuals be capable of there supporting themselves."[17]

Clearly Darwin's "general economy" of life reflected sentiments similar to those expressed in the "political economy" described by Adam Smith. As Gould summed it up, Smith's ideas may not work so well in economics, but they are perfect for biology. And via Smith's insights, Paley's argument for the necessity of a creator is refuted.[18]

"The very phenomena that Paley had revered as the most glorious handiwork of God . . . 'just happen' as a consequence of causes operating at a lower level among struggling individuals," Gould asserted.[19]

THE GAME'S AFOOT

In a way, Darwin's *Origin of Species* represents the third work in a trilogy summarizing the scientific understanding of the world at the end of the 19th century. Just as Newton had tamed the physical world in the 17th century, and Smith had codified economics in the 18th, Charles Darwin in the 19th century added life to the list. Where Smith followed in Newton's footsteps, Darwin followed in Smith's. So by the end of the 19th century, the groundwork was laid for a comprehensive rational understanding of just about everything.

Oddly, it seems, the 20th century produced no such book of similar impact and fame.[20] No volume arrived, for instance, to articulate the long-sought Code of Nature. But one book that appeared in midcentury may someday be remembered as the first significant step toward such a comprehensive handbook of human social behavior: *Theory of Games and Economic Behavior*, by John von Neumann and Oskar Morgenstern.

2

Von Neumann's Games

Game theory's origins

Games combining chance and skill give the best representation of human life, particularly of military affairs and of the practice of medicine which necessarily depend partly on skill and partly on chance. . . . It would be desirable to have a complete study made of games, treated mathematically.

—Gottfried Wilhelm von Leibniz (quoted by Oskar Morgenstern, *Dictionary of the History of Ideas*)

It's no mystery why economics is called the dismal science.

With most sciences, experts make pretty accurate predictions. Mix two known chemicals, and a chemist can tell you ahead of time what you'll get. Ask an astronomer when the next solar eclipse will be, and you'll get the date, time, and best viewing locations, even if the eclipse won't occur for decades.

But mix people with money, and you generally get madness. And no economist really has any idea when you'll see the next total eclipse of the stock market. Yet many economists continue to believe that they will someday practice a sounder science. In fact, some would insist that they are already practicing a sounder science—by viewing the economy as basically just one gigantic game.

At first glance, building economic science on the mathematical theory of games seems about as sensible as forecasting real-estate trends by playing Monopoly. But in the past half century, and particularly the past two decades, game theory has established itself as the precise mathematical tool that economists had long lacked.

Game theory provides precision to the once fuzzy economic notion about how consumers compare their preferences (a measure labeled by the deceptively simple term *utility*). Even more important, game theory shows how to determine the strategies necessary to achieve the maximum possible utility—that is, to acquire the highest payoff—the presumed goal of every rational participant in the dogfights of economic life.

Yet while people have played games for millennia, and have engaged in economic exchange for probably just as long, nobody had ever made the connection explicit—mathematically—until the 20th century. This merger of games with economics—the mathematical mapping of the real world of choices and money onto the contrived realm of poker and chess—has revolutionized the use of math to quantify human behavior. And most of the credit for game theory's invention goes to one of the 20th century's most brilliant thinkers, the magical mathman John von Neumann.

LACK OF FOCUS

If any one person of the previous century personified the word *polymath*, it was von Neumann. I'm really sorry he died so young.

Had von Neumann lived to a reasonably old age—say, 80 or so—I might have had the chance to hear him talk, or maybe even interview him. And that would have given me a chance to observe his remarkable genius for myself. Sadly, he died at the age of 53. But he lived long enough to leave a legendary legacy in several disciplines. His contributions to physics, mathematics, computer science, and economics rank him as one of the all-time intellectual giants of each field. Imagine what he could have accomplished if he'd learned to focus himself!

Of course, he accomplished plenty anyway. Von Neumann produced the standard mathematical formulation of quantum mechanics, for instance. He didn't exactly invent the modern digital computer, but he improved it and pioneered its use for scientific research. And, apparently just for kicks, he revolutionized economics.

Born in 1903 in Hungary, von Neumann was given the name Janos but went by the nickname Jancsi. He was the son of a banker (who had paid for the right to use the honorific title *von*). As a child, Jancsi dazzled adults with his mental powers, telling jokes in Greek and memorizing the numbers in phone books. Later he enrolled in the University of Budapest as a math major, but didn't bother to attend the classes—at the same time, he was majoring in chemistry at the University of Berlin. He traveled back to Budapest for exams, aced them, and continued his chemical education, first at Berlin and then later at the University of Zurich.

I've recounted some of von Neumann's adult intellectual escapades before (in my book *The Bit and the Pendulum*), such as the time when he was called in as a consultant to determine whether the Rand Corporation needed a new computer to solve a difficult problem. Rand didn't need a new computer, von Neumann declared, after solving the problem in his head. In her biography of John Nash, Sylvia Nasar relates another telling von Neumann anecdote, about a famous trick-question math problem. Two cyclists start out 20 miles apart, heading for each other at 10 miles an hour. Meanwhile a fly flies back and forth between the bicycles at 15 miles an hour. How far has the fly flown by the time the bicycles meet? You can solve it by adding up the fly's many shorter and shorter paths between bikes (this would be known in mathematical terms as summing the infinite series). If you detect the trick, though, you can solve the problem in an instant—it will take the bikes an hour to meet, so the fly obviously will have flown 15 miles.

When jokesters posed this question to von Neumann, sure enough, he answered within a second or two. Oh, you knew the trick, they moaned. "What trick?" said von Neumann. "All I did was sum the infinite series."

Before von Neumann first came to America in 1930, he had established himself in Europe as an exceptionally brilliant mathematician, contributing major insights into such topics as logic and set theory, and he lectured at the University of Berlin. But he was not exactly a bookworm. He enjoyed Berlin's cabaret-style nightlife, and more important for science, he enjoyed poker. He turned his talent for both math and cards into a new paradigm for economics—and in so doing devised mathematical tools that someday may reveal deep similarities underlying his many diverse scientific interests. More than that, he showed how to apply rigorous methods to social questions, not unlike Asimov's Hari Seldon.

"Von Neumann was a brilliant mathematician whose contributions to other sciences stem from his belief that impartial rules could be found behind human interaction," writes one commentator. "Accordingly, his work proved crucial in converting mathematics into a key tool to social theory."[1]

UTILITY AND STRATEGY

By most accounts, the invention of modern game theory came in a technical paper published by von Neumann in 1928. But the roots of game theory reach much deeper. After all, games are as old as humankind, and from time to time intelligent thinkers had considered how such games could most effectively be played. As a branch of mathematics, though, game theory did not appear in its modern form until the 20th century, with the merger of two rather simple ideas. The first is utility—a measure of what you want; the second is strategy—how to get what you want.

Utility is basically a measure of value, or preference. It's an idea with a long and complex history, enmeshed in the philosophical doctrine known as utilitarianism. One of the more famous expositors of the idea was Jeremy Bentham, the British social philosopher and legal scholar. Utility, Bentham wrote in 1780, is "that property in any object, whereby it tends to produce benefit, advantage, pleasure, good, or happiness . . . or . . . to prevent the happening of mischief, pain, evil, or unhappiness."[2] So to

Bentham, utility was roughly identical to happiness or pleasure—in "maximizing their utility," individual people would seek to increase pleasure and diminish pain. For society as a whole, maximum utility meant "the greatest happiness of the greatest number."[3] Bentham's utilitarianism incorporated some of the philosophical views of David Hume, friend to Adam Smith. And one of Bentham's influential followers was the British economist David Ricardo, who incorporated the idea of utility into his economic philosophy.

In economics, utility's usefulness depends on expressing it quantitatively. Happiness isn't easily quantifiable, for example, but (as Bentham noted) the *means* to happiness can also be regarded as a measure of utility. Wealth, for example, provides a means of enhancing happiness, and wealth is easier to measure. So in economics, the usual approach is to measure self-interest in terms of money. It's a convenient medium of exchange for comparing the value of different things. But in most walks of life (except perhaps publishing), money isn't everything. So you need a general definition that makes it possible to express utility in a useful mathematical form.

One mathematical approach to quantifying utility came along long before Bentham, in a famous 1738 result from Daniel Bernoulli, the Swiss mathematician (one of many famous Bernoullis of that era). In solving a mathematical paradox about gambling posed by his cousin Nicholas, Daniel realized that utility does not simply equate to quantity. The utility of a certain amount of money, for instance, depends on how much money you already have. A million-dollar lottery prize has less utility for Bill Gates than it would for, say, me. Daniel Bernoulli proposed a method for calculating the reduction in utility as the amount of money increased.[4]

Obviously the idea of utility—what you want to maximize—can sometimes get pretty complicated. But in many ordinary situations, utility is no mystery. If you're playing basketball, you want to score the most points. In chess, you want to checkmate your opponent's king. In poker, you want to win the pot. Often your problem is not defining utility, but choosing a good strategy to maximize it. Game theory is all about figuring out which strategy is best.[5]

The first substantial mathematical attempt to solve that part of the problem seems to have been taken by an Englishman named James Waldegrave in 1713. Waldegrave was analyzing a two-person card game called "le Her," and he described a way to find the best strategy, using what today is known as the "minimax" (or sometimes "minmax") approach. Nobody paid much attention to Waldegrave, though, so his work didn't affect later developments of game theory. Other mathematicians also occasionally dabbled in what is now recognized to be game theory math, but there was no one coherent approach or clear chain of intellectual influence. Only in the 20th century did really serious work begin on devising the mathematical principles behind games of strategy. First was Ernst Zermelo, a German mathematician, whose 1913 paper examining the game of chess is sometimes cited as the beginning of real game theory mathematics. He chose chess merely as an illustration of the more general idea of a two-person game of strategy where the players choose all the moves with no contribution from chance. And that is an important distinction, by the way. Poker involves strategy, but also includes the luck of the draw. If you get a bum hand, you're likely to lose no matter how clever your strategy. In chess, on the other hand, all the moves are chosen by the players—there's no shuffling of cards, tossing of dice, flipping coins, or spinning the wheel of fortune. Zermelo limited himself to games of pure strategy, games without the complications of random factors.

Zermelo's paper on chess apparently confused some of its readers, as many secondary reports of his results are vague and contradictory.[6] But it seems he tried to show that if the White player managed to create an advantageous arrangement of pieces—a "winning configuration"—it would then be possible to end the game within fewer moves than the number of possible chessboard arrangements. (Having an "advantageous arrangement" means achieving a situation from which White would be sure to win—assuming no dumb moves—no matter what Black does.)

Using principles of set theory (one of von Neumann's mathematical specialties, by the way), Zermelo proved that proposition. His original proof required some later tweaking by other mathematicians and Zermelo himself. But the main lesson from it all was not so important for strategy in chess as it was to show that math could be used to analyze important features of any such game of strategy.

As it turns out, chess was a good choice because it is a perfect example of a particularly important type of game of strategy, known as a two-person zero-sum game. It's called "zero-sum" because whatever one player wins, the other loses. The interests of the two competitors are diametrically opposed. (Chess is also a game where the players have "perfect information." That means the game situation and all the decisions of all the players are known at all times—like playing poker with all the cards always dealt face up.)

Zermelo did not address the question of exactly what the best strategy is to play in chess, or even whether there actually is a surefire best strategy. The first move in that direction came from the brilliant French mathematician Émile Borel. In the early 1920s, Borel showed that there is a demonstrable best strategy in two-person zero-sum games—in some special cases. He doubted that it would be possible to prove the existence of a certain best strategy for such games in general.

But that's exactly what von Neumann did. In two-person zero-sum games, he determined, there is always a way to find the best strategy possible, the strategy that will maximize your winnings (or minimize your losses) to whatever extent is possible by the rules of the game and your opponent's choices. That's the modern minimax[7] theorem, which von Neumann first presented in December 1926 to the Göttingen Mathematical Society and then developed fully in his 1928 paper called "Zur Theorie der Gesellshaftsspiele" (Theory of Parlor Games), laying the foundation for von Neumann's economics revolution.[8]

GAMES INVADE ECONOMICS

In his 1928 paper, von Neumann did not attempt to do economics[9]—it was strictly math, proving a theorem about strategic games. Only years later did he merge game theory with economics, with the assistance of an economist named Oskar Morgenstern.

Morgenstern, born in Germany in 1902, taught economics at the University of Vienna from 1929 to 1938. In a book published in 1928, the same year as von Neumann's minimax paper, Morgenstern discussed problems of economic forecasting. A particular point he addressed was the "influence of predictions on predicted events." This, Morgenstern knew, was a problem peculiar to the social sciences, including economics. When a chemist predicts how molecules will react in a test tube, the molecules are oblivious. They do what they do the same way whether a chemist correctly predicts it or not. But in the social sciences, people display much more independence than molecules do. In particular, if people know what you're predicting they will do, they might do something else just to annoy you. More realistically, some people might learn of a prediction and try to turn that foreknowledge to their advantage, upsetting the conditions that led to the prediction and so throwing random factors into the outcome. (By the way, in the Foundation Trilogy, that's why Seldon's Plan had to be so secret. It wouldn't work if anybody knew what it was.)

Anyway, Morgenstern illustrated the problem with a scenario from *The Adventures of Sherlock Holmes*. In the story *The Final Problem*, Holmes was attempting to elude Professor Moriarty while traveling from London to Paris. It wasn't obvious that Holmes could simply outthink Moriarty. Moriarty might anticipate what Holmes was thinking. But then Holmes could anticipate Moriarty's anticipation, and so on: I think that he thinks that I think that he thinks, ad infinitum, or at least nauseum.[10] Consequently, Morgenstern concluded, the situation called for strategy. He returned to the Holmes–Moriarty issue in a 1935 paper exploring the paradoxes of perfect future knowledge.

At that time, after a lecture on these issues, a mathematician named Eduard Čech approached Morgenstern and told him about similar ideas in von Neumann's 1928 paper on parlor games. Morgenstern was entranced, and he awaited an opportunity to meet von Neumann and discuss the relevance of the 1928 paper to Morgenstern's views on economics.

The chance came in 1938, when Morgenstern accepted a three-year appointment to lecture at Princeton University. (Von Neumann had by then taken up his position at the nearby Institute for Advanced Study.) "The principal reason for my wanting to go to Princeton," Morgenstern said, "was the possibility that I might become acquainted with von Neumann."[11] As Morgenstern told the story, he soon revived von Neumann's interest in game theory and began writing a paper to show its relevance to economics. As von Neumann critiqued early drafts, the paper grew longer, with von Neumann eventually joining Morgenstern as a coauthor. By this time—it was now 1940—the paper had grown substantially, and it kept growing, ultimately into a book published by the Princeton University Press in 1944. (Subsequent historical study suggests, though, that von Neumann had previously written most of the book without Morgenstern's help.[12])

Theory of Games and Economic Behavior instantly became the game theory bible. In the eyes of game theory believers, it was to economics what Newton's *Principia* was to physics. It was a sort of newtonizing of Adam Smith, providing mathematical rigor to describe how individual interactions affect a collective economy. "We hope to establish," wrote von Neumann and Morgenstern, "that the typical problems of economic behavior become strictly identical with the mathematical notions of suitable games of strategy." It will become apparent, they asserted, that "this theory of games of strategy is the proper instrument with which to develop a theory of economic behavior."[13] The authors then developed the theory throughout more than 600 pages, dense with equations and diagrams. But the opening sections are remarkably readable, laying out the authors' goals and intentions in a kind of extended preamble designed to persuade skeptical economists that their field needed an overhaul.

While noting that many economists had already been using mathematics, von Neumann and Morgenstern declared that "its use has not been highly successful," especially when compared to other sciences such as physics. Throughout its early pages, the book draws on physics as the model for how math can make murky knowledge precise and practical—in contrast to economics, where the basic ideas had been expressed so fuzzily that past efforts to use math had been doomed. "Economic problems . . . are often stated in such vague terms as to make mathematical treatment a priori appear hopeless because it is quite uncertain what the problems really are," the authors wrote.[14] What economics needed was a theory that made precise and meaningful measurements possible, and game theory filled the bill.

Von Neumann and Morgenstern were careful to emphasize, though, that their theory was just a first step. "There exists at present no universal system of economic theory," they wrote, and if such a theory were ever to be developed, "it will very probably not be during our lifetime."[15] But game theory could provide the foundation for such a theory, by focusing on the simplest of economic interactions as a guide to developing general principles that would someday be able to solve more complicated problems. Just as modern physics began when Galileo studied the rather simple problem of falling bodies, economics could benefit from a similar understanding of simple economic behavior.

"The great progress in every science came when, in the study of problems which were modest as compared with ultimate aims, methods were developed that could be extended further and further," von Neumann and Morgenstern declared.[16] And so it made sense to focus on the simplest aspect of economics—the economic interaction of individual buyers and sellers. While economic science as a whole involves the entire complicated system of producing and pricing goods, and earning and spending money, at the root of it all is the choicemaking of the individuals participating in the economy.

ROBINSON CRUSOE MEETS GILLIGAN

Back in the days when von Neumann and Morgenstern were working all this out, standard economic textbooks extolled a simple economic model of their own, called the "Robinson Crusoe" economy. Stranded on a desert island, Crusoe was an economy unto himself. He made choices about how to use the resources available to him to maximize his utility, coping only with the circumstances established by nature.

Samuel Bowles, an economist at the University of Massachusetts, explained to me that textbooks viewed economics as just the activities of many individual Robinson Crusoes. Where Crusoe interacted with nature, consumers in a big-time economy interacted with prices. And that was the standard "neoclassical" view of economic theory. "That's what everybody taught," Bowles said. "But there was something odd about it." It seemed to be a theory of social interactions based on someone who had interacted only nonsocially, that is, with nature, not with other people. "Game theory adopts a different framework," Bowles said. "I'm in a situation in which my well-being depends on what somebody else does, and your well-being depends on what I do—therefore we are going to think strategically."[17]

And that's exactly the point that von Neumann and Morgenstern stressed back in 1944. The Robinson Crusoe economy is fundamentally different, conceptually, from a Gilligan's Island economy. It's not just the complication of social influences from other people affecting your choices about the prices of goods and services. The results of your choices—and thus your ability to achieve your desired utility—are inevitably intertwined with the choices of the others. "If two or more persons exchange goods with each other, then the result for each one will depend in general not merely upon his own actions but on those of the others as well," von Neumann and Morgenstern declared.[18]

Mathematically, that meant that no longer could you simply compute a single simple maximum utility for Robinson Crusoe. Your calculations had to accommodate a mixture of competing goals, maximum utilities for Gilligan, the Skipper too, the millionaire, and his wife, the movie star, the Professor and Mary Ann. "This kind of problem is nowhere dealt with in classical mathematics," von Neumann and Morgenstern noted.

Indeed, Bentham's notion of the "greatest possible good of the greatest possible number" is mathematically meaningless. It's like saying you want the most possible food at the least possible cost. Think about it—you can have zero cost (and no food) or all the food in the world, at a very high cost. Which do you want? You certainly can't calculate an answer to that question. In a Gilligan's Island economy, it's not really an issue of wanting the maximum utility for the maximum number, but rather that all the individuals want their own personal possible maximum. In other words, "All maxima are desired at once—by various participants."[19] And in trying to fulfill their desires, every individual's actions will be influenced by expectations of everyone else's actions, and vice versa, the old "I think he thinks I think" problem. That makes a social economy, with multiple participants, inherently distinct from the Robinson Crusoe economy. "And it is this problem which the theory of 'games of strategy' is mainly devised to meet," von Neumann and Morgenstern announced.[20]

Of course, it is easier said than done. It's one thing to realize that Gilligan's Island is more complex than Robinson Crusoe's; it's something else again to figure out how to do the math. Sure, you can start with something simple, like analyzing the interactions between just two people. Then, once you understand how two people will interact, you can use the same principles to analyze what will happen when a third person enters the game, and then a fourth, and so on. (Eventually, then, you would possess the elusive Code of Nature, once you mastered the math of analyzing the behavior of all the individuals in society as a whole.)

However, you can see how things would rapidly become difficult to keep track of. Each person in the game (or the economy) will make choices based on a wide range of variables. In the Robinson Crusoe economy, his set of variables encompasses all those factors that would affect his quest for maximum utility. But if the *Minnow* had beached on Crusoe's island, each new player would have brought an additional set of variables of his or her own into the game. Then Crusoe would need to take all of those new variables into account, too.

On top of all that, more players means a more complex economy, more kinds of goods and services, different methods of production. So the social economy rapidly becomes a mathematical nightmare, it would seem, beyond even the ability of the know-it-all Professor to resolve. But there is hope, for economics and for understanding society, and it's a hope that's based on the simple idea of taking a temperature.

TAKING SOCIETY'S TEMPERATURE

In drawing analogies between economics and physics, von Neumann and Morgenstern talked a lot about the theory of heat (or, as it is more pretentiously known, thermodynamics). They pointed out, for instance, that measuring heat precisely did not lead to a theory of heat; physicists needed the theory first, in order to understand how to measure heat in an unambiguous way. In a similar way, game theory needed to be developed first to give economists the tools they needed to measure economic variables properly.

The example of the theory of heat played another crucial role—in articulating a basic issue within game theory itself. At the outset, von Neumann and Morgenstern made it clear that they did not want to venture into the philosophical quagmire of defining all the nuances of utility. For them, to develop game theory for use in economics, it was enough to equate utility with money. For the businessman, money (as in profits) is a logical measure of utility; for consumers, income (minus expenses) is a good measure of utility, or you could think of the utility of an object as the price you were willing to pay for it. And money can be used as a currency for translating what anyone wants into more specific objects or events or experiences or whatever. So equating utility with money is a convenient simplifying assumption, allowing the theory to focus on the strategic aspects of how to achieve what you want, without worrying about the complications involved in defining what you want.

However, there remained an important aspect of utility that von Neumann and Morgenstern had to address. Was it even possible, in the first place, to define utility in a numerical way, to make it susceptible to a mathematical theory? (Bernoulli had proposed a way to calculate utility, but he had not tried to prove that the concept could be a basis for making rational choices in a consistent way.) Money (which obviously is numerical) could really be a good stand-in for the more complex concept of utility only if utility can really be represented by a numerical concept. And so they had to show that it was possible to define utility in a mathematically rigorous way. That meant identifying axioms from which the notion of utility could be deduced and measured quantitatively.

As it turned out, utility could be quantified in a way not unlike the approach physicists used to construct a scientifically rigorous definition of temperature. After all, primitive notions of utility and temperature are similar. Utility, or preference, can be thought of as just a rank ordering. If you prefer A to B, and B to C, you surely prefer A to C. But it is not so obvious that you can ascribe a number to *how much* you prefer A to B, or B to C. It was once much the same with heat—you could say that something felt warmer or cooler than something else, but not necessarily how much, certainly not in a precise way—before the development of the theory of heat. But nowadays the absolute temperature scale, based on the laws of thermodynamics, gives temperature an exact quantitative meaning. And von Neumann and Morgenstern showed how you could similarly convert rank orderings into numerically precise measures of utility.

You can get the essence of the method from playing a modified version of *Let's Make a Deal*. (For the youngsters among you, that was a famous TV game show, in which host Monty Hall offered contestants a chance to trade their prizes for possibly more valuable prizes, at the risk of getting a clunker.) Suppose Monty offers you three choices: a BMW convertible, a top-of-the-line big-screen plasma TV, or a used tricycle. Let's say you want the BMW most of all, and that you'd prefer the TV to the tricycle. So it's a simple matter to rank the relative utility of the three products. But here comes the deal. Your choice is to get either the plasma TV, OR a 50-50 chance of getting the BMW. That is, the TV is behind Door Number 1, and the BMW is behind either Door Number 2 or Door Number 3. The other door conceals the tricycle.

Now you really have to think. If you choose Door Number 1—the plasma TV—you must value it at more than 50 percent as much as the BMW. But suppose the game is more complicated, with more doors, and the odds change to a 60 percent chance of the BMW, or 70 percent. At some point you will be likely to opt for the chance to get the BMW, and at that point, you could conclude that the utilities are numerically equal—you value the plasma TV at, say, 75 percent as much as the BMW (plus 25 percent of the tricycle, to be technically precise). Consequently, to give utility a numerical value, you just have to arbitrarily assign some number to one choice, and then you can compare other choices to that one using the probabilistic version of *Let's Make a Deal*.

So far so good. But there remains the problem of operating in a social economy where your personal utility is not the only issue—you have to anticipate the choices of others. And in a small-scale Gilligan's Island economy, pure strategic choices can be subverted by things like coalitions among some of the players. Again, the theory of heat offers hope.

Temperature is a measure of how fast molecules are moving. In principle, it's not too hard to describe the velocity of a single molecule, just as you could easily calculate Robinson Crusoe's utility. But you'd have a hard time with Gilligan's Island, just as it becomes virtually impossible to keep track of all the speeds of a relatively few number of interacting molecules. But if you have a trillion trillion molecules or so, the interactions tend to average out, and using the theory of heat you can make precise predictions about temperature. (The math behind this is, of course, statistical mechanics, which will become even more central to the game theory story in later chapters.)

As von Neumann and Morgenstern pointed out, "very great numbers are often easier to handle than those of medium size."[21] That was exactly the point made by Asimov's psychohistorians: Even though you can't track each individual molecule, you can predict the aggregate behavior of vast numbers, precisely what taking the temperature of a gas is all about. You can measure a value related to the average velocity of all the molecules, which reflects the way the individual molecules interact. Why not do the same for people? It worked for Hari Seldon. And it might work for a sufficiently large economy. "When the number of participants becomes really great," von Neumann and Morgenstern wrote, "some hope emerges that the influence of every particular participant will become negligible."[22]

With the basis for utility established at the outset, von Neumann and Morgenstern could proceed simply by taking money to be utility's measure. The bulk of their book was then devoted to the issue of finding the best strategy to make the most money.

At this point, it's important to clarify what they meant by the concept of strategy. A strategy in game theory is a very specific course of action, not a general approach to the game. It's not like tennis, for instance, where your strategy might be "play aggressively" or "play safe shots." A game theory strategy is a defined set of choices to make for every possible circumstance that might arise. In tennis, your strategy might be to "never rush the net when your opponent serves; serve and volley whenever you are even or ahead in a game; always stay back when behind in a game." And you'd have other rules for all the other situations.

There's one additional essential point about strategy in game theory—the distinction between "pure" strategies and "mixed" strategies. In tennis, you might rush the net after every serve (a pure strategy) or you might rush the net after one out of every three serves, staying back at the baseline two times out of three (a mixed strategy). Mixed strategies often turn out to be essential for making game theory work.

In any event, the question isn't whether there is always a good *general* strategy, but whether there is always an optimum set of rules for strategic behavior that covers all eventualities. And in fact, there is—for two-person zero-sum games. You can find the best strategy using the minimax theorem that von Neumann published in 1928. His proof of that theorem was notoriously complicated. But its essence can be boiled down into something fairly easy to remember: When playing poker, sometimes you need to bluff.

MASTERING MINIMAX

The secret behind the minimax approach in two-person zero-sum games is the need to remember that whatever one player wins, the other loses (the definition of zero sum). So your strategy should seek to maximize your winnings, which would have the effect of minimizing your opponent's winnings. And of course your opponent wants to do the same.

Depending on the game, you may be able to play as well as possible and still not win anything, of course. The rules and stakes may be such that whoever plays first will always win, for instance, and if you go second, you're screwed. Still, it is likely that some strategies will lose more than others, so you would attempt to minimize your opponent's gains (and your losses). The question is, what strategy do you choose to do so? And should you stick with that strategy every time you play?

It turns out that in some games, you may indeed find one pure strategy that will maximize your winnings (or minimize your losses) no matter what the other player does. Obviously, then, you would play that strategy, and if the game is repeated, you would play the same strategy every time. But sometimes, depending on the rules of the game, your wisest choice will depend on what your opponent does, and you might not know what that choice will be. That's where game theory gets interesting.

Let's look at an easy example first. Say that Bob owes Alice $10. Bob proposes a game whereby if he wins, his debt will be reduced. (In the real world, Alice will tell Bob to take a hike and fork over the $10.) But for purposes of illustrating game theory, she might agree.

Bob suggests these rules: He and Alice will meet at the library. If he gets there first, he pays Alice $4; if she gets there first, he pays her $6. If they both arrive at the same time, he pays $5. (As I said, Alice would probably tell him to shove it.)

Now, let's say they live together, or at least live next door to each other. They both have two possible strategies for getting to the library—walking or taking the bus. (They are too poor to own a car, which is why Bob is haggling over the $10.) And they both know that the bus will always beat walking. So this game is trivial—both will take the bus, both will arrive at the same time, and Bob will pay Alice $5.[23] And here's how game theory shows what strategy to choose: a "payoff matrix." The numbers show how much the player on the left (Alice) wins.

		Bob	
		Bus	Walk
Alice	Bus	5	6
	Walk	4	5

In a zero-sum game, the numbers in a payoff matrix designate how much the person on the left (in this case, Alice) wins (since it's zero sum, the numbers tell how much the player on top, Bob, loses). If the number is negative, that means the player on top wins that much (negative numbers signaling a loss for Alice). In non-zero-sum games, each matrix cell will include two numbers, one for each player (or more if there are more players, which makes it very hard to show the matrix for multiplayer games).

Obviously, Alice must choose the bus strategy because it always does as well as or better than walking, no matter what Bob does. And Bob will choose the bus also, because it minimizes his losses, no matter what Alice does. Walking can do no better and might be worse.

Of course, you didn't need game theory to figure this out. So let's look at another example, from real-world warfare, a favorite of game theory textbooks.

In World War II, General George Kenney knew that the Japanese would be sending a convoy of supply ships to New Guinea. The Allies naturally wanted to bomb the hell out of the convoy. But the convoy would be taking one of two possible routes—one to the north of New Britain, one to the south.

Either route would take three days, so in principle the Allies could get in three days' worth of bombing time against the convoy. But the weather could interfere. Forecasters said the northern route would be rainy one of the days, limiting the bombing time to a maximum of two days. The southern route would be clear, providing visibility for three days of bombing. General Kenney had to decide whether to send his reconnaissance planes north or south. If he sent them south and the convoy went north, he would lose a day of bombing time (of only two bombing days available). If the recon planes went north, the bombers would still have time to get two bombing days in if the convoy went south.

So the "payoff" matrix looks like this, with the numbers giving the Allies' "winnings" in days of bombing:

		Japanese	
		North	South
Allies	North	2	2
	South	1	3

If you just look at this game matrix from the Allies' point of view, you might not see instantly what the obvious strategy is. But from the Japanese side, you can easily see that going north is the only move that makes sense. If the convoy took the southern route, it was guaranteed to get bombed for two days and maybe even three. By going north, it would get a maximum of two days (and maybe only one), as good as or better than any of the possibilities going south. General Kenney could therefore confidently conclude that the Japanese would go north, so the only logical Allies strategy would be to send the reconnaissance planes north as well. (The Japanese did in fact take the northern route and suffered heavy losses from the Allied bombers.)

Proper strategies are not, of course, always so obvious. Let's revisit Alice and Bob and see what happened after Alice refused to play Bob's stupid game. Knowing that she was unlikely ever to get her whole $10 back, she proposed another game that would cause Bob to scratch his head about what strategy to play.

In Alice's version of the game, they go the library every weekday for a month. If they both ride the bus, Bob pays Alice $3. If they both walk, Bob pays Alice $4. If Bob rides the bus and Alice walks, arriving second, Bob pays $5. If Bob walks and Alice rides the bus, arriving first, Bob pays $6. If you are puzzled, don't worry. This game puzzles Bob, too. Here's the matrix.

		Bob	
		Bus	Walk
Alice	Bus	3	6
	Walk	5	4

Bob realizes there is no simple strategy for playing this game. If he rides the bus, he might get off paying only $3. But Alice, realizing that, will probably walk, meaning Bob would have to pay her $5.

So Bob might decide to walk, hoping to pay only $4. But then Alice might figure that out and ride the bus, so Bob would have to pay her $6. Neither Bob nor Alice can be sure of what the other will do, so there is no obvious "best" strategy.

Remember, however, that Alice required the game to be played repeatedly, say for a total of 20 times. Nothing in the rules says you have to play the same strategy every day. (If you did, that would be a pure strategy—one that never varied.) To the contrary, Alice realizes that she should play a mixed strategy—some days walking, some days riding the bus. She wants to keep Bob guessing. Of course, Bob wants to keep Alice guessing too. So he will take a mixed strategy approach also.

And that was the essence of von Neumann's ingenious insight. In a two-person zero-sum game, you can *always* find a best strategy—it's just that in many cases the best strategy is a mixed strategy.

In this particular example, it's easy to calculate the best mixed strategies for Alice and Bob. Remember, a mixed strategy is a mix of pure strategies, each to be chosen a specific percentage of the time (or in other words, with a specific probability).[24] So Bob wants to compute the ratio of the percentages for choosing "walk" versus "bus," using a recipe from an old game theory book that he found in the library.[25] Following the book's advice, he compares the payoffs for each choice when Alice walks (the first row of the matrix) to the values when Alice takes the bus (the second row of the matrix), subtracting the payoffs in the second row from those in the first. (The answers are -2 and 2, but the minus sign is irrelevant.) Those two numbers determine the best ratio for Bob's two strategies—2:2, or 50-50. (Note, however, that it is the number in the second column that determines the proportion for the first strategy, and the number in the first column that determines the proportion for the second strategy. It just so happened that in this case the numbers are equal.) For Alice, on the other hand, subtracting the second column from the first column gives -3 and 1 (or 3 and 1, ignoring the minus sign). So she should play the first strategy (bus) three times as often as the second strategy (walk).[26]

Consequently, Alice should ride the bus one time in four and walk three-fourths of the time. Bob should ride the bus half the time and walk half the time. Both should decide which strategy to choose by using some suitable random-choice device. Bob could just flip a coin; Alice might use a random number table, or a game spinner with three-fourths of the pie allocated to walking and one-fourth to the bus.[27] If either Alice or Bob always walked (or took the bus), the other would be able to play a more profitable strategy.

So you have to keep your opponent guessing. And that's why game theory boils down to the need to bluff while playing poker. If you always raise when dealt a good hand but never when dealt a poor hand, your opponents will be able to figure out what kind of a hand you hold.

Real poker is too complicated for an easy game theory analysis. But consider a simple two-player version of poker, where Bob and Alice are each dealt a single card, and black always beats red.[28] Before the cards are dealt, each player antes $5, so there is $10 in the pot. Alice then plays first, and she may either fold or bet an additional $3. If she folds, both players turn over their cards, and whoever holds a black card wins the pot. (If both have black or both have red, they split the pot.)

If Alice wagers the additional $3, Bob can then either match the $3 and call (making a total of $16 in the pot) or fold. If he folds, Alice takes the $13 in the pot; if he calls, they turn over their cards to see who wins the $16.

You'd think, at first, that if Alice had a red card she'd simply pass and hope that Bob also had red. But if she bets, Bob might think she must have black. If he has red, he might fold—and Alice will win with a red card. Bluffing sometimes pays off. On the other hand, Bob knows that Alice might be bluffing (since she is not a Vulcan), and so he may go ahead and call.

The question is, how often should Alice bluff, and how often should Bob call her (possible) bluff? Maybe von Neumann could have figured that out in his head, but I think most people would need game theory.

A matrix for this game would show that both players can choose from four strategies. Alice can always pass, always bet, pass with red and bet with black, or bet with red and pass with black. Bob can always fold, always call, fold with red and call with black, or fold with black and call with red. If you calculate the payoffs, you will see that Alice should bet three-fifths of the time no matter what card she has; the other two-fifths of the time she should bet only if she has black. Bob, on the other hand, should call Alice's bet two-fifths of the time no matter what card he has; three-fifths of the time he should fold if he has red and call if he has black.[29] (By the way, another thing game theory can show you is that this game is stacked in favor of Alice, if she always goes first. Playing the mixed strategies dictated by the game matrix assures her an average of 30 cents per hand.)

The notion of a mixed strategy, using some random method to choose from among the various pure strategies, is the essence of von Neumann's proof of the minimax theorem. By choosing the correct mixed strategy, you can guarantee the best possible outcome you can get—if your opponent plays as well as possible. If your opponent doesn't know game theory, you might do even better.

BEYOND GAMES

Game theory was not supposed to be just about playing poker or chess, or even just about economics. It was about making strategic decisions—whether in the economy or in any other realm of real life. Whenever people compete or interact in pursuit of some goal, game theory describes the outcomes to be expected by the use of different strategies. If you know what outcome you want, game theory dictates the proper strategy for achieving it. If you believe that people interacting with other people are all trying to find the best possible strategy for achieving their desires, it makes sense that game theory might potentially be relevant to the modern idea of a Code of Nature, the guide to human behavior.

In their book, von Neumann and Morgenstern did not speak

of a "Code of Nature," but did allude to game theory as a description of "order of society" or "standard of behavior" in a social organization. And they emphasized how a "theory of social phenomena" would require a different sort of math from that commonly used in physics—such as the math of game theory. "The mathematical theory of games of strategy," they wrote, "gains definitely in plausibility by the correspondence which exists between its concepts and those of social organizations."[30]

In its original form, though, game theory was rather limited as a tool for coping with real-world strategic problems. You can find examples of two-person zero-sum games in real life, but they are typically either so simple that you don't need game theory to tell you what to do, or so complicated that game theory can't incorporate all the considerations.

Of course, expecting the book that introduces a new field to solve all of that field's problems would be a little unrealistic. So it's no surprise that in applying game theory to situations more complicated than the two-person zero-sum game, von Neumann and Morgenstern were not entirely successful. But it wasn't long before game theory's power was substantially enhanced, thanks to the beautiful math of John Forbes Nash.

3

Nash's Equilibrium

Game theory's foundation

Nash's theory of noncooperative games should now be recognized as one of the outstanding intellectual advances of the twentieth century . . . comparable to that of the discovery of the DNA double helix in the biological sciences.

—economist Roger Myerson

As letters of recommendation go, it was not very elaborate, just a single sentence: "This man is a genius."

That was how Carnegie Tech professor R. L. Duffin described John Nash to the faculty at Princeton University, where Nash entered as a 20-year-old graduate student in 1948. Within two years, Duffin's assessment had been verified. Nash's "beautiful mind" had by then launched an intellectual revolution that eventually propelled game theory from the fad du jour to the foundation of the social sciences.

Shortly before Nash's arrival at Princeton, von Neumann and Morgenstern had opened a whole new continent for mathematical exploration with the groundbreaking book *Theory of Games and Economic Behavior*. It was the Louisiana Purchase of economics. Nash played the role of Lewis and Clark.

As it turned out, Nash spent more time in the wilderness than Lewis and Clark did, as mental illness robbed the rationality of the

man whose math captured rationality's essence. But before his pro-
longed departure, Nash successfully steered game theory toward
the mathematical equivalent of manifest destiny. Though not
warmly welcomed at first, Nash's approach to game theory eventually captured a major share of the economic-theory market, leading to his Nobel Prize for economics in 1994. By then game theory
had also conquered evolutionary biology and invaded political science, psychology, and sociology. Since Nash's Nobel, game theory
has infiltrated anthropology and neuroscience, and even physics.
There is no doubt that game theory's wide application throughout
the intellectual world was made possible by Nash's math.

"Nash carried social science into a new world where a unified
analytical structure can be found for studying all situations of conflict and cooperation," writes University of Chicago economist
Roger Myerson. "The theory of noncooperative games that
Nash founded has developed into a practical calculus of incentives
that can help us to better understand the problems of conflict
and cooperation in virtually any social, political, or economic
institution."[1]

So it's not too outrageous to suggest that in a very real way,
Nash's math provides the foundation for a modern-day Code of
Nature. But of course it's not as simple as that. Since its inception,
game theory has had a complicated and controversial history. Today it is worshiped by some but still ridiculed by others. Some
experimenters claim that their results refute game theory; others
say the experiments expand game theory and refine it. In any event,
game theory has assumed such a prominent role in so many realms
of science that it can no longer intelligently be ignored, as it often
was in its early days.

IGNORED AT BIRTH

When von Neumann and Morgenstern introduced game theory as
the math for economics, it made quite a splash. But most economists remained dry. In the mid-1960s, the economics guru Paul
Samuelson praised the von Neumann–Morgenstern book's insight

and impact—in other fields. "The book has accomplished every-thing except what it started out to do—namely, revolutionize economic theory," Samuelson wrote.[2]

It's not that economists hadn't heard about it. In the years following its publication, *Theory of Games and Economic Behavior* was widely reviewed in social science and economics journals. In the *American Economic Review*, for example, Leonid Hurwicz admired the book's "audacity of vision" and "depth of thought."[3] "The potentialities of von Neumann's and Morgenstern's new approach seem tremendous and may, one hopes, lead to revamping, and enriching in realism, a good deal of economic theory," Hurwicz wrote. "But to a large extent they are only potentialities: results are still largely a matter of future developments."[4] A more enthusiastic assessment appeared in a mathematics journal, where a reviewer wrote that "posterity may regard this book as one of the major scientific achievements of the first half of the twentieth century."[5]

The world at large also soon learned about game theory. In 1946, the von Neumann–Morgenstern book rated a front page story in the *New York Times*; three years later a major piece appeared in *Fortune* magazine.

It was also clearly appreciated from the beginning that game theory promised applications outside economics—that (as von Neumann and Morgenstern had themselves emphasized) it contained elements of the long-sought theory of human behavior generally. "The techniques applied by the authors in tackling economic problems are of sufficient generality to be valid in political science, sociology, or even military strategy," Hurwicz pointed out in his review.[6] And Herbert Simon, a Nobel laureate-to-be, made similar observations in the *American Journal of Sociology*. "The student of the Theory of Games . . . will come away from the volume with a wealth of ideas for application . . . of the theory into a fundamental tool of analysis for the social sciences."[7]

Yet it was also clear from the outset that the original theory of games was severely limited. Von Neumann had mastered two-person zero-sum games, but introducing multiple players led to problems. Game theory worked just fine if Robinson Crusoe was playing games with Friday, but the math for Gilligan's Island wasn't as rigorous.

Von Neumann's approach to multiple-player games was to assume that coalitions would form. If Gilligan, the Skipper, and Mary Ann teamed up against the Professor, the Howells, and Ginger, you could go back to the simple two-person game rules. Many players might be involved, but if they formed two teams, the teams could take the place of individual players in the mathematical analysis.

But as later commentators noted, von Neumann had led himself into an inconsistency, threatening his theory's internal integrity. A key part of two-person zero-sum games was choosing a strategy that was the best you could do against a smart opponent. Your best bet was to play your optimal (probably mixed) strategy *no matter what anybody else did.* But if coalitions formed among players in many-person games, as von Neumann believed they would, that meant your strategy would in fact depend on coordinating it with at least some of the other players. In any event, game theory describing many players interacting in non-zero-sum situations— that is, game theory applicable to real life—needed something more than the original *Theory of Games* had to offer. And that's what John Nash provided.

BEAUTIFUL MATH

The book *A Beautiful Mind* offers limited insight into Nash's math, particularly in regard to all the many areas of science where that math has lately become prominent.[8] But the book reveals a lot about Nash's personal troubles. Sylvia Nasar's portrait of Nash is not very flattering, though. He is depicted as immature, self-centered, arrogant, uncaring, and oblivious. But brilliant.

Nash was born in West Virginia, in the coal-mining town of Bluefield, in 1928. While showing some interest in math in high school (he even took some advanced courses at a local college), he planned to become an electrical engineer, like his father. But by the time he enrolled at Carnegie Tech (the Carnegie Institute of Technology) in Pittsburgh, his choice for major had become chemical engineering. He soon switched to chemistry, but that didn't last, either. Finding no joy in manipulating laboratory apparatus, Nash turned to math, where he excelled.

He first mixed math with economics while taking an undergraduate course at Carnegie Tech in international economics. In that class Nash conceived the idea for a paper on what came to be called the "bargaining problem." As later observers noted, it was a paper obviously written by a teenager—not because it was intellectually naive, but because the bargaining he considered was over things like balls, bats, and pocket knives. Nevertheless the mathematical principles involved were clearly relevant to more sophisticated economic situations.

When Nash arrived at Princeton in 1948, it had already become game theory's world capital. Von Neumann was at the Institute for Advanced Study, just a mile from the university, and Morgenstern was in the Princeton economics department. And at the university math department, a cadre of young game theory enthusiasts had begun exploring the new von Neumann–Morgenstern continent in earnest. Nash himself attended a game theory seminar led by Albert W. Tucker (but also explored game theory's implications on his own).

Shortly after his arrival, Nash realized that his undergraduate ideas about the "bargaining problem" represented a major new game theory insight. He prepared a paper for publication (with assistance from von Neumann and Morgenstern, who "gave helpful advice as to the presentation").

"Bargaining" represents a different form of game theory in which the players share some common concerns. Unlike the two-person zero-sum game, in which the loser loses what the winner wins, a bargaining game offers possible benefits to both sides. In this "cooperative" game theory, the goal is for all players to do the best they can, but not necessarily at the expense of the others. In a good bargain, both sides gain. A typical real-life bargaining situation would be negotiations between a corporation and a labor union.

In his bargaining paper, Nash discussed the situation when there is more than one way for the players to achieve a mutual benefit. The problem is to find which way maximizes the benefit (or utility) for both sides—given that both players are rational (and know how to quantify their desires), are equally skilled bargainers, and are equally knowledgeable about each other's desires.

When bargaining over a possible exchange of resources (in Nash's example, things like a book, ball, pen, knife, bat, and hat), the two players might assess the values of the objects differently. (To the athletic minded, a bat might seem more valuable than a book, while the more intellectually oriented bargainer might rank the book more valuable than the bat.) Nash showed how to consider such valuations and compute each player's gain in utility for various exchanges, providing a mathematical map for finding the location of the optimal bargain—the one giving the best deal for both (in terms of maximizing the increase in their respective utilities).[9]

SEEKING EQUILIBRIUM

Nash's bargaining problem paper would in itself have established him as one of game theory's leading pioneers. But it was another paper, soon to become his doctoral dissertation, that established Nash as the theory's prophet. It was the paper introducing the "Nash equilibrium," eventually to become game theory's most prominent pillar.

The idea of equilibrium is, of course, immensely important to many realms of science. Equilibrium means things are in balance, or stable. And stability turns out to be an essential idea for understanding many natural processes. Biological systems, chemical and physical systems, even social systems all seek stability. So identifying how stability is reached is often the key to predicting the future. If a situation is unstable—as many often are—you can predict the future course of events by figuring out what needs to happen to achieve stability. Understanding stability is a way of knowing where things are going.

The simplest example is a rock balanced atop a sharply peaked hill. It's not a very stable situation, and you can predict the future pretty confidently: That rock is going to roll down the hill, reaching an equilibrium point in the valley. Another common example of equilibrium shows up when you try to dissolve too much sugar in a glass of iced tea. A pile of sugar will settle at the bottom of the glass. When the solution reaches equilibrium, molecules will continue to dissolve out of the pile, but at the same rate as other sugar molecules drop out of the tea and join the pile. The tea is then in a stable situation, maintaining a constant sweetness.

It's the same principle, just a little more complicated, in a chemical reaction, where stability means achieving a state of "chemical equilibrium," in which the amounts of the reacting chemicals and their products remain constant. In a typical reaction, two different chemical substances interact to produce a new, third substance. But it's often not the case that both original substances will entirely disappear, leaving only the new one. At first, amounts of the reacting substances will diminish as the quantity of the product grows. But eventually you may reach a point where the amount of each substance doesn't change. The reaction continues—but as the first two substances react to make the third, some of the third decomposes to replenish supplies of the first two. In other words, the action continues, but the big picture doesn't change.

That's chemical equilibrium, and it is described mathematically by what chemists call the law of mass action. Nash had just this sort of physical equilibrium in mind when he was contemplating stability in game theory. In his dissertation he refers to "the 'mass-action' interpretation of equilibrium," and that such an equilibrium is approached in a game as players "accumulate empirical information" about the payoffs of their strategies.[10]

When equilibrium is reached in a chemical reaction, the quantities of the chemicals no longer change; when equilibrium is reached in a game, nobody has any incentive to change strategies—so the choice of strategies should remain constant (the game situation is, in other words, stable). All the players should be satisfied with the strategy they've adopted, in the sense that no other strategy would do better (as long as nobody else changes strategies, either). Similarly, in social situations, stability means that everybody is pretty much content with the status quo. It may not be that you like things the way they are, but changing them will only make things worse. There's no impetus for change, so like a rock in a valley, the situation is at an equilibrium point.

In a two-person zero-sum game, you can determine the equilibrium point using von Neumann's minimax solution. Whether using a pure strategy or a mixed strategy, neither player has anything to gain by deviating from the optimum strategy that game theory prescribes. But von Neumann did not prove that similarly stable solutions could be found when you moved from the Robinson Crusoe–Friday economy to the Gilligan's Island economy or Manhattan Island economy. And as you'll recall, von Neumann thought the way to analyze large economies (or games) was by considering coalitions among the players.

Nash, however, took a different approach—deviating from the "party line" in game theory, as he described it decades later. Suppose there are no coalitions, no cooperation among the players. Every player wants the best deal he or she can get. Is there always a set of strategies that makes the game stable, giving each player the best possible personal payoff (assuming everybody chooses the best available strategy)? Nash answered yes. Borrowing a clever piece of mathematical trickery known as a "fixed-point theorem," he proved that every game of many players (as long as you didn't have an infinite number of players) had an equilibrium point.

Nash derived his proof in different ways, using either of two fixed-point theorems—one by Luitzen Brouwer, the other by Shizuo Kakutani. A detailed explanation of fixed-point theorems requires some dense mathematics, but the essential idea can be illustrated rather simply. Take two identical sheets of paper, crumple one up, and place it on top of the other. Somewhere in the crumpled sheet will be a point lying directly above the corresponding point on the uncrumpled sheet. That's the fixed point. If you don't believe it, take a map of the United States and place it on the floor—any floor within the United States. (The map represents the crumpled up piece of paper.) No matter where you place the map, there will be one point that is directly above the corresponding actual location in the United States. Applying the same principle to the players in a game, Nash showed that there was always at least one "stable" point where competing players' strategies would be at equilibrium.

"An equilibrium point," he wrote in his Ph.D. thesis, "is . . . such that each player's mixed strategy maximizes his payoff if the strategies of the others are held fixed."[11] In other words, if you're playing such a game, there is at least one combination of strategies such that if you change yours (and nobody else changes theirs) you'll do worse. To put it more colloquially, says economist Robert Weber, you could say that "the Nash equilibrium tells us what we might expect to see in a world where no one does anything wrong."[12] Or as Samuel Bowles described it to me, the Nash equilibrium "is a situation in which everybody is doing the best they can, given what everybody else is doing."[13]

Von Neumann was dismissive of Nash's result, as it did turn game theory in a different direction. But eventually many others recognized its brilliance and usefulness. "The concept of the Nash equilibrium is probably the single most fundamental concept in game theory," declared Bowles. "It's absolutely fundamental."[14]

GAME THEORY GROWS UP

Nash published his equilibrium idea quickly. A brief (two-page) version appeared in 1950 in the *Proceedings of the National Academy of Sciences*. Titled "Equilibrium Points in *n*-Person Games," the paper established concisely (although not particularly clearly for nonmathematicians) the existence of "solutions" to many-player games (a solution being a set of strategies such that no single player could expect to do any better by unilaterally trying a different strategy). He expanded the original paper into his Ph.D. thesis, and a longer version was published in 1951 in *Annals of Mathematics*, titled "Non-cooperative Games."

Von Neumann and Morgenstern, Nash politely noted in his paper, had produced a "very fruitful" theory of two-person zero-sum games. Their theory of many-player games, however, was restricted to games that Nash termed "cooperative," in the sense that it analyzed the interactions among coalitions of players. "Our theory, in contradistinction, is based on the *absence* of coalitions in that it is assumed that each participant acts independently, without collaboration or communication with any of the others."[15] In other words, Nash devised an "every man for himself" version of many-player games—which is why he called it "noncooperative" game theory. When you think about it, that approach pretty much sums up many social situations. In a dog-eat-dog world, the Nash equilibrium describes how every dog can have its best possible day. "The distinction between non-cooperative and cooperative games that Nash made is decisive to this day," wrote game theorist Harold Kuhn.[16]

To me, the really key point about the Nash equilibrium is that it cements the analogy between game theory math and the laws of physics—game theory describing social systems, the laws of physics describing natural systems. In the natural world, everything seeks stability, which means seeking a state of minimum energy. The rock rolls downhill because a rock at the top of a hill has a high potential energy; it gives that energy away by rolling downhill. It's because of the law of gravity. In a chemical reaction, all the atoms involved are seeking a stable arrangement, possessing a minimum amount of energy. It's because of the laws of thermodynamics.

And just as in a chemical reaction all the atoms are simultaneously seeking a state with minimum energy, in an economy all the people are seeking to maximize their utility. A chemical reaction reaches an equilibrium enforced by the laws of thermodynamics; an economy should reach a Nash equilibrium dictated by game theory.[17]

Real life isn't quite that simple, of course. There are usually complicating factors. A bulldozer can push the rock back up the hill; you can add chemicals to spark new chemistry in a batch of molecules. When people are involved, all sorts of new sources of unpredictability complicate the game theory playing field. (Imagine how much trickier chemistry would become if molecules could think.[18])

Nevertheless, Nash's notion of equilibrium captures a critical feature of the social world. Using Nash's math, you can figure out how people could reach stability in a social situation by comparing that situation to an appropriate game. So if you want to apply game theory to real life, you need to devise a game that captures the essential features of the real-life situation you're interested in. And life, if you haven't noticed, poses all sorts of different circumstances to cope with.

Consequently game theorists have invented more games than you can buy at Toys Я Us. Peruse the game theory literature, and you'll find the matching pennies game, the game of chicken, public goods games, and the battle of the sexes. There's the stag hunt game, the ultimatum game, and the "so long sucker" game. And hundreds of others. But by far the most famous of all such games is a diabolical scenario known as the Prisoner's Dilemma.

TO RAT OR NOT TO RAT

As in all my books, a key point has once again been anticipated by Edgar Allan Poe. In "The Mystery of Marie Roget," Poe described a murder believed by Detective Dupin to have been committed by a gang. Dupin's strategy is to offer immunity to the first member of the gang to come forward and confess. "Each one of a gang, so placed, is not so much . . . anxious for escape, as fearful of betrayal," Poe's detective reasoned. "He betrays eagerly and early that he may not himself be betrayed."[19] It's too bad that Poe (who was in fact a trained mathematician) had not thought to work out the math of betrayal—he might have invented game theory a century early.

As it happened, the Prisoner's Dilemma in game theory was first described by Nash's Princeton professor, Albert W. Tucker, in 1950. At that time, Tucker was visiting Stanford and had mentioned his game theory interests. He was asked unexpectedly to present a seminar, so he quickly conjured up the scenario of two criminals captured by the police and separately interrogated.[20]

You know the story. The police have enough evidence to convict two criminal conspirators on a lesser offense, but need one or the other to rat out his accomplice to make an armed robbery charge stick. So if both keep mum, both will get a year in prison. But whoever agrees to testify goes free. If only one squeals, the partner gets five years. If both sing like a canary, then both get three years (a two-year reduction for copping a plea).

| | | Alice | |
		Keep Mum	Rat
Bob	Keep Mum	1, 1	5, 0
	Rat	0, 5	3, 3

Years in prison for Bob, Alice

If you look at this game matrix, you can easily see where the Nash equilibrium is. There's only one combination of choices where neither player has any incentive to change strategies—they both rat each other out. Think about it. Let's say our game theory experts Alice and Bob have decided to turn to crime, but the police catch them. The police shine a light in Bob's face and spell out the terms of the game. He has to decide right away. He ponders what Alice might do. If Alice rats him out—a distinct possibility, knowing Alice—his best choice is to rat her out, too, thereby getting only three years instead of five. But suppose Alice keeps mum. Then Bob's best choice is still to rat her out, as he'll then get off free. *No matter which strategy Alice chooses, Bob's best choice is betrayal,* just as Poe's detective had intuited. And Alice, obviously, must reason the same way about Bob. The only stable outcome is for both to agree to testify, ratting out their accomplice.

Ironically, and the reason it's called a dilemma, they would both be better off overall if they both kept quiet. But they are interrogated separately, with no communication between them permitted. So the best strategy for each individual produces a result that is not the best result for the team. If they both keep mum (that is, they cooperate with each other), they spend a total of two years in prison (one each). If one rats out the other (technical term: defects), but the other keeps mum, they serve a total of five years (all by the silent partner). But when they rat each other out, they serve a total of six years—a worse overall outcome than any of the other pairs of strategies. The Nash equilibrium—the stable pair of choices dictated by self-interest—produces a poorer payoff for the group. From the standpoint of game theory and Nash's math, the choice is clear. If everybody's incentive is to get the best individual deal, the proper choice is to defect.

In real life, of course, you never know what will happen, because the crooks may have additional considerations (such as the prospect of sleeping with the fishes if they rat out the wrong guy). Consequently the Nash equilibrium calculation does not always predict how people will really behave. Sometimes people temper their choices with considerations of fairness, and sometimes they act out of spite. In Prisoner's Dilemma situations, some people actually do choose to cooperate. But that doesn't detract from the importance of the Nash equilibrium, as economists Charles Holt and Alvin Roth point out. "The Nash equilibrium is useful not just when it is itself an accurate predictor of how people will behave in a game but also when it is not," they write, "because then it identifies a situation in which there is a tension between individual incentives and other motivations." So if people cooperate (at least at first) in a Prisoner's Dilemma situation, Nash's math tells us that such cooperation, "because it is not an equilibrium, is going to be unstable in ways that can make cooperation difficult to maintain."[21]

Though it is a simplified representation of real life, the Prisoner's Dilemma game does capture the essence of many social interactions. But obviously you cannot easily assess any social situation by calculating the Nash equilibrium. Real-life games often involve many players and complicated payoff rules. While Nash showed that there is always at least one equilibrium point, it's another matter to figure out what that point is. (And often there is more than one Nash equilibrium point, which makes things really messy.) Remember, each player's "strategy" will typically be a mixed strategy, drawn from maybe dozens or hundreds or thousands (or more) of pure "specific" strategies. In most games with many players, calculating all the probabilities for all the combinations of all those choices exceeds the computational capacity of Intel, Microsoft, IBM, and Apple put together.

THE PUBLIC GOOD

It's not hopeless, though. Consider another favorite game to illustrate "defection"—the public goods game. The idea is that some members in a community reap the benefits of membership without paying their dues. It's like watching public television but never calling in to make a pledge during the fund drives. At first glance, the defector wins this game—getting the benefit of enjoying Morse and Poirot without paying a price. But wait a minute. If everybody defected, there would be no benefit for anybody. The free riders would become hapless hitchhikers.

Similarly, suppose your neighborhood association decided to collect donations to create a park. You'd enjoy the park, but if you reason that enough others in the neighborhood will contribute enough money to build it, you might decline to contribute. If everybody reasons the same way, though, there will be no park. But suppose that defecting (declining to pay) and cooperating (contributing your fair share) are not the only possible strategies. You can imagine a third strategy, called reciprocating. If you are a reciprocator, you pay only if you know that a certain number of the other players have decided to pay. Computer simulations of this kind of game suggest that a mix of these strategies among the players can reach a Nash equilibrium.

Experiments with real people show the same thing. One study, reported in 2005, tested college students on a contrived version of the public goods game. Four players were each given tokens (representing money) and told they could contribute as many as they liked into a "public pot," keeping the rest in their personal account. The experimenter then doubled the number of tokens in the pot. One player at a time was told how much had been contributed to the pot and then given a chance to change his or her contribution. When the game ended (after a random number of rounds), all the tokens were then evenly divided up among all the players.

How would you play? Since, in the end, all four players split the pot equally, the people who put in the least to begin with end up with the most tokens—their share of the pot *plus* the money they held back in their personal account. Of course, if nobody put any in to begin with, nobody reaped the benefit of the experimenter's largesse, kind of like a local government forgoing federal matching funds for a highway project. So it would seem to be a good strategy to donate *something* to the pot. But if you want to get a better payoff than anyone else, you should put in less than the others. Maybe one token. On the other hand, everybody in the group will get more if you put more in the pot to begin with. (That way, you might not get more than everybody else, but you'll get more than you otherwise would.)

When groups of four played this game repeatedly, a pattern of behavior emerged. Players fell into three readily identifiable groups: cooperators, defectors (or "free riders"), and reciprocators. Since all the players learned at some point how much had been contributed, they could adjust their behavior accordingly. Some players remained stingy (defectors), some continued to contribute generously (cooperators), and others contributed more *if* others in the group had donated significantly (reciprocators).

Over time, the members of each group earned equal amounts of money, suggesting that something like a Nash equilibrium had been achieved—they all won as much as they could, given the strategy of the others. In other words, in this kind of game, the human race plays a mixed strategy—about 13 percent cooperators,

20 percent defectors (free riders), and 63 percent reciprocators in this particular experiment. "Our results support the view that our human subject population is in a stable . . . equilibrium of types," wrote the researchers, Robert Kurzban and Daniel Houser.[22] Knowing about the Nash equilibrium helps make sense of results like these.

GAME THEORY TODAY

Together with his paper on the bargaining problem (which treats cooperative game situations), Nash's work on equilibria in many-player games greatly expanded game theory's scope beyond von Neumann and Morgenstern's book, providing the foundation for much of the work in game theory going on today. There's more to game theory than the Nash equilibrium, of course, but it is still at the heart of current endeavors to apply game theory to society broadly.

Over the years, game theorists have developed math for games where coalitions do form, where information is incomplete, where players are less than perfectly rational. Models of all of these situations, plus many others, can be built using game theory's complex mathematical tools. It would take a whole book (actually, several books) to describe all of those subsequent developments (and many such books have been written). It's not necessary to know all those details of game theory history, but it is important to know that game theory does have a rich and complex history. It is a deep and complicated subject, full of many highly technical and nuanced contributions of substantial mathematical sophistication.

Even today game theory remains very much a work in progress. Many deep questions about it do not seem to have been given compelling answers. In fact, if you peruse the various accounts of game theory, you are likely to come away confused. Its practitioners do not all agree on how to interpret some aspects of game theory, and they certainly disagree about how to advertise it.

Some presentations seem to suggest that game theory should predict human behavior—what choices people will make in games (or in economics or other realms of life). Others insist that game theory does not predict, but prescribes—it tells you what you ought to do (if you want to win the game), not what any player would actually do in a game. Or some experts will say that game theory predicts what a "rational" person will do, acknowledging that there's no accounting for how irrational some people (even those playing high-stakes games) can be. Of course, if you ask such experts to define "rational," they're likely to say that it means behaving in the way that game theory predicts.

To me, it seems obvious that basic game theory does not always successfully predict what people will do, since most people are about as rational as pi. Neither is it obvious that game theory offers a foolproof way to determine what *is* the rational thing to do. There may always be additional considerations in making a "rational" choice that have not been included in game theory's mathematical framework.

Game theory *does* predict outcomes for different strategies in different situations, though. In principle you could use game theory to analyze lots of ordinary games, like checkers, as well as many problems in the real world where the concept of game is much broader. It can range from trying to beat another car into a parking place to global thermonuclear war. The idea is that when faced with deciding what to do in some strategic interaction, the math can tell you which move is most likely to be successful. So if you know what you want to achieve, game theory can help you—if your circumstances lend themselves to game theory representation.

The question is, are there ever any such circumstances? Early euphoria about game theory's potential to illuminate social issues soon dissipated, as a famous game theory text noted in 1957. "Initially there was a naive band-wagon feeling that game theory solved innumerable problems of sociology and economics, or that, at the least, it made their solution a practical matter of a few years' work. This has not turned out to be the case."[23]

Such an early pessimistic assessment isn't so surprising. There's always a lack of patience in the scientific world; many people want new ideas to pay off quickly, even when more rational observers realize that decades of difficult work may be needed for a theory to reach maturity. But even six decades after the von Neumann–Morgenstern book appeared, you could find some rather negative assessments of game theory's relevance to real life.

In an afterword to the 60-year-anniversary edition of *Theory of Games*, Ariel Rubenstein acknowledged that game theory had successfully entrenched itself in economic science. "Game theory has moved from the fringe of economics into its mainstream," he wrote. "The distinction between economic theorist and game theorist has virtually disappeared."[24] But he was not impressed with claims that game theory was really good for much else, not even games. "Game theory is not a box of magic tricks that can help us play games more successfully. There are very few insights from game theory that would improve one's game of chess or poker," Rubenstein wrote.[25]

He scoffed at theorists who believed game theory could actually predict behavior, or even improve performance in real-life strategic interactions. "I have never been persuaded that there is a solid foundation for this belief," he wrote. "The fact that the academics have a vested interest in it makes it even less credible." Game theory in Rubenstein's view is much like logic—form without substance, a guide for comparing contingencies but not a handbook for action. "Game theory does not tell us which action is preferable or predict what other people will do. . . . The challenges facing the world today are far too complex to be captured by any matrix game."[26]

OK—maybe this book should end here. But no. I think Rubenstein has a point, but also that he is taking a very narrow view. In fact, I think his attitude neglects an important fact about the nature of science.

Scientists make models. Models capture the essence of some aspect of something, hopefully the aspect of interest for some particular use or another. Game theory is all about making models of human interactions. *Of course* game theory does not capture all the nuances of human behavior—no model does. No map of Los Angeles shows every building, every crack in every sidewalk, or every pothole—if it showed all that, it wouldn't be a map of Los Angeles, it would *be* Los Angeles. Nevertheless, a map that leaves out all those things can still help you get where you want to go (although in L.A. you might get there slowly).

Naturally, game theory introduces simplifications—it is, after all, a *model* of real-life situations, not real life itself. In that respect it is just like all other science, providing simplified models of reality that are accurate enough to draw useful conclusions about that reality. You don't have to worry about the chemical composition of the moon and sun when predicting eclipses, only their masses and motions. It's like predicting the weather. The atmosphere is a physical system, but Isaac Newton was no meteorologist. Eighteenth-century scholars did not throw away Newton's *Principia* because it couldn't predict thunderstorms. But after a few centuries, physics *did* get to the point where it could offer reasonably decent weather forecasts. Just because game theory cannot predict human behavior infallibly today doesn't mean that its insights are worthless.

In his book *Behavioral Game Theory*, Colin Camerer addresses these issues with exceptional insight and eloquence. It is true, he notes, that many experiments produce results that seem—at first—to disconfirm game theory's predictions. But it's clearly a mistake to think that therefore there is something wrong with game theory's math. "If people don't play the way theory says, their behavior does not prove the mathematics wrong, any more than finding that cashiers sometimes give the wrong change disproves arithmetic," Camerer points out.[27] Besides, game theory (in its original form) is based on players' behaving rationally and selfishly. If actual real-life behavior departs from game theory's forecast, perhaps there's just something wrong with the concepts of rationality and selfishness. In that case, incorporating better knowledge of human psychology (especially in social situations) into game theory's equations can dramatically improve predictions of human behavior and help explain why that behavior is sometimes surprising. That is exactly the sort of thing that Camerer's specialty, behavioral game theory, is intended to do. "The goal is not to '*dis*prove' game theory . . . but to *im*prove it," Camerer writes.[28]

As it turns out, game theory *is* widely used today in scientific efforts to understand all sorts of things. While Nash's 1994 Nobel Prize recognized the math establishing game theory's foundations, the 2005 economics Nobel trumpeted the achievements of two important pioneers of game theory's many important applications. Economist Thomas Schelling, of the University of Maryland, understood in the 1950s that game theory offered a mathematical language suitable for unifying the social sciences, a vision he articulated in his 1960 book *The Strategy of Conflict.* "Schelling's work prompted new developments in game theory and accelerated its use and application throughout the social sciences," the Royal Swedish Academy of Sciences remarked on awarding the prize.[29]

Schelling paid particular attention to game-theoretic analysis of international relations, specifically (not surprising for the time) focusing on the risks of armed conflict. In gamelike conflict situations with more than one Nash equilibrium, Schelling showed how to determine which of the equilibrium possibilities was most plausible. And he identified various counterintuitive conclusions about conflict strategy that game theory revealed. An advancing general burning bridges behind him would seem to be limiting his army's options, for example. But the signal sent to the enemy—that the oncoming army had no way to retreat—would likely diminish the opposition's willingness to fight. Similar reasoning transferred to the economic realm, where a company might decide to build a big, expensive production plant, even if it meant a higher cost of making its product, if by flaunting such a major commitment it scared competitors out of the market.

Schelling's insights also extended to games where all the players desire a common (coordinated) outcome more than any particular outcome—in other words, when it is better for everybody to be on the same page, regardless of what the page is. A simple example would be a team of people desiring to eat dinner at the same restaurant. It doesn't matter what restaurant (as long as the food is not too spicy); the goal is for everyone to be together. When everybody can communicate with each other, coordination is rarely a problem (or at least it shouldn't be), but in many such situations communication is restricted. Schelling shed considerable light on the game-theoretic issues involved in reaching coordinated solutions to such social problems. Some of Schelling's later work applied game theory to the rapid change in some neighborhoods from a mixture of races to being largely segregated, and to limits on individual control over behavior—why people do so many things they really don't want to do, like smoke or drink too much, while not doing things they really want to, like exercising.

2005's other economics Nobel winner, Robert Aumann, has long been a leading force in expanding the scope of game theory to many disciplines, from biology to mathematics. A German-born Israeli at the Hebrew University of Jerusalem, Aumann took special interest in long-term cooperative behavior, a topic of special relevance to the social sciences (after all, long-term cooperation is the defining feature of civilization itself). In particular, Aumann analyzed the Prisoner's Dilemma game from the perspective of infinitely repeated play, rather than the one-shot deal in which both players' best move is to rat the other out. Over the long run, Aumann showed, cooperative behavior can be sustained even by players who still have their own self-interest at heart.

Aumann's "repeated games" approach had wide application, both in cases where it led to cooperation and where it didn't. By showing how game theory's rules could facilitate cooperation, he also identified the circumstances where cooperation was less likely—when many players are involved, for instance, or when communication is limited or time is short. Game theory helps to show why certain common forms of collective behavior materialize under such circumstances. "The repeated-games approach clarifies the raison d'être of many institutions, ranging from merchant guilds and organized crime to wage negotiations and international trade agreements," the Swedish academy pointed out.

While Nobel Prizes shine the media spotlight on specific achievements of game theory, they tell only a small portion of the whole story. Game theory's uses have expanded to multiple arenas in recent years. Economics is full of applications, from guiding negotiations between labor unions and management to auctioning licenses for exploiting the electromagnetic spectrum. Game theory is helpful in matching medical residents to hospitals, in understanding the spread of disease, and in determining how to best vaccinate against various diseases—even to explain the incentives (or lack thereof) for hospitals to invest in fighting bacterial resistance to antibiotics. Game theory is valuable for understanding terrorist organizations and forecasting terrorist strategies. For analyzing voting behaviors, for understanding consciousness and artificial intelligence, for solving problems in ecology, for comprehending cancer. You can call on game theory to explain why the numbers of male and female births are roughly equal, why people get stingier as they get older, and why people like to gossip about other people.

Gossip, in fact, turns out to be a crucial outcome of game theory in action, for it's at the heart of understanding human social behavior, the Code of Nature that made it possible for civilization to establish itself out of the selfish struggles to survive in the jungle. For it is in biology that game theory has demonstrated its power most dramatically, in explaining otherwise mysterious outcomes of Darwinian evolution. After all, people may not always play game theory the way you'd expect, but animals do, where the Code of Nature really *is* the law of the jungle.

4

Smith's Strategies

Evolution, altruism, and cooperation

The stunning variety of life forms that surround us, as well as the beliefs, practices, techniques, and behavioral forms that constitute human culture, are the product of evolutionary dynamics.

—Herbert Gintis, *Game Theory Evolving*

To understand human sociality we have much to learn from primates, birds, termites, and even dung beetles and pond scum.

—Herbert Gintis, *Game Theory Evolving*

In the winter of 1979, Cambridge University biologist David Harper decided it would be fun to feed the ducks.

A flock of 33 mallards inhabited the university's botanical garden, hanging out at a particular pond where they foraged for food. Daily foraging is important for ducks, as they must maintain a minimum weight for low-stress flying. Unlike landlubber animals that can gorge themselves in the fall and live off their fat in the winter, ducks have to be prepared for takeoff at any time. They therefore ought to be good at finding food fast, in order to maintain an eat-as-you-go lifestyle.

Harper wanted to find out just how clever the ducks could be at maximizing their food intake. So he cut up some white bread into precisely weighed pieces and enlisted some friends to toss the pieces onto the pond.

The ducks, naturally, were delighted with this experiment, so they all rapidly paddled into position. But then Harper's helpers began tossing the bread onto two separated patches of the pond. At one spot, the bread tosser dispensed one piece of bread every five seconds. The second was slower, tossing out the bread balls just once every 10 seconds.

Now, the burning scientific question was, if you're a duck, what do you do? Do you swim to the spot in front of the fast tosser or the slow tosser? It's not an easy question. When I ask people what they would do, I inevitably get a mix of answers (and some keep changing their mind as they think about it longer).

Perhaps (if you were a duck) your first thought would be to go for the guy throwing the bread the fastest. But all the other ducks might have the same idea. You'd get more bread for yourself if you switched to the other guy, right? But you're probably not the only duck who would realize that. So the choice of the optimum strategy isn't immediately obvious, even for people. To get the answer you have to calculate a Nash equilibrium.

After all, foraging for food is a lot like a game. In this case, the chunks of bread are the payoff. You want to get as much as you can. So do all the other ducks. As these were university ducks, they were no doubt aware that there is a Nash equilibrium point, an arrangement that gets every duck the most food possible when all the other ducks are also pursuing a maximum food-getting strategy.

Knowing (or observing) the rate of tosses, you can calculate the equilibrium point using Nash's math. In this case the calculation is pretty simple: The ducks all get their best possible deal if one-third of them stand in front of the slow tosser and the other two-thirds stand in front of the fast tosser.

And guess what? It took the ducks about a minute to figure that out. They split into two groups almost precisely the size that game theory predicted. *Ducks know how to play game theory!*

When the experimenters complicated things—by throwing bread chunks of different sizes—the ducks needed to consider both the rate of tossing and the amount of bread per toss. Even then, the ducks eventually sorted themselves into the group sizes that Nash equilibrium required, although it took a little longer.[1]

Now you have to admit, that's a little strange. Game theory was designed to describe how "rational" humans would maximize their utility. And now it turns out you don't need to be rational, or even human.[2] The duck experiment shows, I think, that there's more to game theory than meets the eye. Game theory is not just a clever way to figure out how to play poker. Game theory captures something about how the world works.

At least the biological world. And it was in fact the realization that game theory describes biology that gave it its first major scientific successes. Game theory, it turns out, captures many features of biological evolution. Many experts believe that it explains the mystery of human cooperation, how civilization itself could emerge from individuals observing the laws of the jungle. And it even seems to help explain the origin of language, including why people like to gossip.

LIFE AND MATH

I learned about evolution and game theory by visiting the Institute of Advanced Study in Princeton, home of von Neumann during game theory's infancy. Long recognized as one of the world's premier centers for math and physics, the institute had been slow to acknowledge the ascent of biology in the hierarchy of scientific disciplines. By the late 1990s, though, the institute had decided to plunge into the 21st century a little early by initiating a program in theoretical biology.

Just as the newborn institute had reached across the Atlantic to bring von Neumann, Einstein, and others to America, it recruited a director for its biology program from Europe—Martin Nowak, an Austrian working at the University of Oxford in England. Nowak was an accomplished mathematical biologist who had mixed biochemistry with math during his student years at the University of Vienna, where he earned his doctorate in 1988. He soon moved on to Oxford, where he eventually became head of the mathematical biology program. I visited him in Princeton in the fall of 1998 to inquire about the institute's plans for mixing math with the science of life.

Nowak described a diverse research program, touching on everything from the immune system—deciphering the math behind fighting the AIDS virus, for instance—to inferring the origins of human language. Underlying much of his work was a common theme that at the time I really didn't appreciate: the pervasive relevance of game theory.

It makes sense, of course. In biology almost everything involves interaction. The sexes interact to reproduce, obviously. There are the fierce interactions of immune system cells battling viruses, or toxic molecules tangling with DNA to cause cancer. And humans, of course, always interact—cooperatively or contentiously, or just by talking to each other.

Evolutionary processes shape the way that such interactions occur and what their outcomes will be. And that's a key point: Evolution is not just about the origin of new species from common ancestors. Evolution is about virtually everything in biology— the physiology of individuals, the diversity of appearances within groups, the distribution of species in an ecosystem, and the behavior of individuals in response to other individuals or groups interacting with other groups. Evolution underlies all the biological action, and underlying evolution's power is the mathematics of game theory. "Game theory has been very successfully used in evolution," Nowak told me. "An overwhelming number of problems in evolution are of a game-theoretic nature."[3]

In particular, game theory helps explain the evolution of social behavior in the animal (including humans) kingdom, solving a perplexing mystery in the original formulation of Darwinism: Why do animals cooperate? You'd think that the struggle to survive would put a premium on selfishness. Yet cooperation is common in the biological world, from symbiotic relationships between parasites and their hosts to out-and-out altruism that people often exhibit toward total strangers. Human civilization could never have developed as it has without such widespread cooperation; finding the Code of Nature describing human social behavior will not be possible without understanding how that cooperation evolved. And the key clues to that understanding are coming from game theory.

GAMES OF LIFE

In the 1960s, even before most economists took game theory seriously, several biologists noticed that it might prove useful in explaining aspects of evolution. But the man who really put evolutionary game theory on the scientific map was the British biologist John Maynard Smith.

He was "an approachable man with unruly white hair and thick glasses," one of his obituaries noted, "remembered by colleagues and friends as a charismatic speaker but deadly debater, a lover of nature and an avid gardener, and a man who enjoyed nothing better than discussing scientific ideas with young researchers over a glass of beer in a pub."[4] Unfortunately I never had a chance to have a beer with him. He died in 2004.

Maynard Smith was born in 1920. As a child, he enjoyed collecting beetles and bird-watching, foreshadowing his future biological interests. At Eton College he was immersed in mathematics and then specialized in engineering at Cambridge University. During World War II he did engineering research on airplane stability, but after the war he returned to biology, studying zoology under the famed J. B. S. Haldane at University College London.

In the early 1970s, Maynard Smith received a paper to review that had been submitted to the journal *Nature* by an American researcher named George Price. Price had attempted to explain why animals competing for resources did not always fight as ferociously as they might have, a puzzling observation if natural selection really implied that they should fight to the death if only the fittest survive. Price's paper was too long for *Nature*, but the issue remained in the back of Maynard Smith's mind. A year later, while visiting the theoretical biology department at the University of Chicago, he studied game theory and began to explore the ways in which evolution is like a game.[5]

Eventually, Maynard Smith showed that game theory could illuminate how organisms adopt different strategies to survive the slings and arrows of ecological fortune and produce offspring to carry the battle on to future generations. Evolution is a game that all life plays. All animals participate; so do plants, so do bacteria. You don't need to attribute any rationality or reasoning power to the organisms—their strategy is simply the sum of their properties and propensities. Is it a better strategy to be a short tree or a tall tree? To be a super speedy quadruped or a slower but smarter biped? Animals don't choose their strategies so much as they *are* their strategies.

This is a curious observation, I think. If every animal (plant, bug) is a different strategy, then why are there so many different forms of life out there, why so many different strategies for surviving? Why isn't there one best strategy? Why doesn't one outperform all the others, making it the sole survivor, the winner of the ultimate fitness sweepstakes? Darwin, of course, had dealt with that issue, explaining how different kinds of survival advantages could be exploited by natural selection to diversify life into a smorgasbord of species (like the specialization of workers in Adam Smith's pin factory). Maynard Smith, though, took the Darwinian explanation to greater depths, using game theory to demonstrate with mathematical rigor why evolution is not a winner-takes-all game.

In doing so, Maynard Smith perceived the need to modify classical game theory in two ways: substituting the evolutionary ideas of "fitness" for utility and "natural selection" for rationality. In economic game theory, he noted, "utility" is somewhat artificial; it's a notion that attempts "to place on a single linear scale a set of qualitatively distinct outcomes" such as a thousand dollars, "losing one's girl friend, losing one's life." In biology, though, "fitness, or expected number of offspring, may be difficult to measure, but it is unambiguous. There is only one correct way of combining different components—for example, chances of survival and of re-production."[6] And "rationality" as a strategy for human game play-ers exhibits two "snags," Maynard Smith noted: "It is hard to decide what is rational, and in any case people do not behave rationally." Consequently, he asserted, "the effect of these changes is to make game theory more readily applicable in biology than in the human sciences."[7]

To illustrate his insight, he invented a clever but simple animal-fighting game. Known as the hawk-dove game, it showed why one single strategy would not produce a stable population. Imagine such a world, a "bird planet" populated solely by birds. These birds are capable of behaving either like hawks (aggressive, always ready to fight over food), or doves (always peaceful and passive). Now suppose these birds all decide that being hawkish is the best survival strategy. Whenever two of them encounter some food, they fight over it—the winner eats, the loser nurses his wounds, starves, and maybe even dies. But even the winner may suffer some injuries, incurring a cost that diminishes its benefits from getting the food.

Now suppose one of these hawkish birds decides that all this fighting is . . . well, for the birds. He starts behaving like a dove. Upon encountering some food, he eats only if no other bird is around. If one of those hawks shows up, the "dove" flies away. The dove might miss a few meals, but at least he's not losing his feathers in fights. Furthermore, suppose a few other birds try the dove approach. When they meet each other, they share the food. While the hawks are chewing each other up, the doves are chew-ing on dinner.

Consequently, Maynard Smith noted, an all-hawk population is not an "evolutionary stable strategy." An all-hawk society is sus-ceptible to invasion by doves. On the other hand, it is equally true that an all-dove society is not stable, either. The first hawk who comes along will eat pretty well, because all the other birds will fly away at the sight of him. Only when more hawks begin to appear will there be any danger of dying in a fight. So the question is, what *is* the best strategy? Hawk or dove?

It turns out that the best strategy for surviving depends on how many hawks there are in the population. If hawks are rare, a hawkish strategy is best because most of the opponents will be doves and will run from a fight. If hawks are plentiful, though, they will get into many costly fights—yielding an advantage for dovish behavior. So a society should evolve to include a mix of hawks and doves. The higher the cost of fighting, the fewer the number of hawks. Maynard Smith showed how game theory described this situation perfectly, with an evolutionary stable strategy being the biological counterpart of a Nash equilibrium.

While an evolutionary stable strategy is analogous to a Nash equilibrium, it is not always precisely equivalent. In many sorts of games there can be more than one Nash equilibrium, and some of them may not be evolutionary stable strategies. An ecosystem composed of various species with a fixed set of behavioral strategies could be at a Nash equilibrium without being immune to invasion by a mutant capable of introducing a new strategy into the competition. Such an ecosystem would not be evolutionarily stable.[8] But the birds are unlikely to appreciate that distinction. In any case, the birds have to choose to play hawk or dove just as the ducks had to decide which bread tosser to favor. The best mix—the evolutionary stable strategy—will be a split population, some percentage doves, some hawks.

Exactly what those percentages are depends on the precise costs of fighting compared to the food you miss by fleeing. Here's one game matrix showing a possible weighting of the costs:

		Bird 2	
		Hawk	Dove
Bird 1	Hawk	−2, −2	2, 0
	Dove	0, 2	1, 1

If two hawks meet, both are losers (getting "scores" of −2) because they beat each other up. If Bird 1 is a hawk and Bird 2 a dove, the dove flies away and gets 0, the hawk gets all the food (2). But if two doves meet, they share the food and both get 1 point. (Or you could say that one dove defers to the other half the time, the 1 point each signifying a 50-50 chance of either bird getting the food.) If you calculate it out, you find that the best mix of strategies (for these values of the costs) is that two-thirds should be doves and one-third hawks.[9] (Keep in mind that, mathematically, you could have a mix of hawks and doves, or just birds that play mixed strategies. In other words, if you're a bird in this scenario, your best bet is to behave like a hawk one-third of the time and behave like a dove two-thirds of the time.)[10]

Obviously this is a rather simplified view of biology. Hawks and doves are not the only possible behavioral strategies, even for birds. But you can see the basic idea, and you should also be able to see how game theory could describe situations with added complexity.

Suppose, for instance, "spectator birds" watched as other birds battled. In fact, like human boxing or football fans, some birds do like to watch the gladiators of their group slug it out in a good fight (as do certain fishes). And that desire to view violence may offer a clue to why societies provide so much violence to view. Spectating may be wired into animal genes by evolutionary history, and maybe game theory has something to do with it.

At first glance, spectating offers one obvious survival advantage—you're less likely to get killed watching than fighting. But you don't have to be a spectator to avoid the danger of a fight. You can simply get as far away from any fighting as you can. So why watch? The answer emerges naturally from game theory. You may find yourself in an unavoidable fight someday, in which case it would be a good idea to know your opponent's record.

Face it: You can't always run from a fight. The wimps who retreat from every encounter don't really enhance their chance of survival, for they will lose out in the competition for food, mates, and other essential resources. On the other hand, looking for a fight at every opportunity is not so smart, either—the battle may exact a greater cost than the benefit of acquiring the resource. You would expect clever birds to realize that they might have to fight someday, so they better scout their potential opponents by observing them in battle. The observers (or "eavesdroppers" in biolingo) could choose to be either a hawk or a dove when it's their turn to fight—depending on what they've observed about their adversary.

Rufus Johnstone, of the University of Cambridge, extended the math of the hawk-dove game in just this manner to evaluate the eavesdropper factor. In this game, the eavesdropper knows whether its opponent has won or lost its previous fight. An eavesdropper encountering a loser will act hawkish, but if encountering a winner the eavesdropper will adopt a dove strategy and forgo the chance to win the resource.

"An individual that is victorious in one round is more likely to win in the next, because its opponent is less likely to mount an escalated challenge," Johnstone concluded.[11]

Since eavesdroppers have the advantage of knowing when to run, avoiding fights with dangerous foes, you might guess that eavesdropping would reduce the amount of violent conflict in a society. Alas, the math shows otherwise. Adding eavesdroppers to the hawk-dove game raises the rate of "escalated" fighting—occasions where both combatants take the hawk approach.

Why? Because of the presence of spectators! If nobody is watching, it is not so bad to be a dove. But in the jungle, reputation is everything. With spectators around, acting like a dove guarantees that you'll face an aggressive opponent in your next fight. Whereas if everybody sees that you're a ferocious hawk, your next opponent may head for the hills at the sight of you.

So the presence of spectators encourages violence, and watching violence today offers an advantage for the spectators who may be fighters tomorrow. In other words, the benefit to an individual of eavesdropping—helping that individual avoid high-risk conflict—drives a tendency toward a higher level of high-risk conflict in the society as a whole.

But don't forget that adding spectators is just one of many

complications that could be considered in the still very simplified hawk-dove game. Fights depend on more than just aggressiveness. Size and skill come into play as well. And one study noted that a bird's self-assessment of its own fighting skills can also influence the fight-or-flight decision. If the birds know their own skill levels accurately, overall fighting might be diminished. (You can think of this as the Clint Eastwood version of the hawk-dove game: A bird has got to know its limitations.)[12]

In any case, policy makers who would feel justified in advocating wars based on game theory should pause and realize that real life is more complicated than biologists' mathematical games. Humans, after all, have supposedly advanced to a civilized state where the law of the jungle doesn't call all the shots. And in fact, game theory can help show how that civilized state came about. Game theory describes how the circumstances can arise that make cooperation and communication a stable strategy for the members of a species. Without game theory, cooperative human social behavior is hard to understand.

EVOLVING ON A LANDSCAPE

Game theory can help illuminate how different strategies fare in the battle to survive. Even more important, game theory helps to show how the best strategies might differ as circumstances change. After all, a set of behavioral propensities that's successful in the jungle might not be such a hot idea in the Antarctic.

When evolutionists talk about circumstances changing, typically they'll be referring to something like the climate, or the trauma of a recent asteroid impact. But the changing strategies of the organisms themselves can be just as important. And that's why game theory is essential for understanding evolution. Remember the basic concept of a Nash equilibrium—it's when everybody is doing the best they can do, given what everybody else is doing. In other words, the best survival strategy depends on who else is around and how they are behaving. If your survival hinges on the actions of others, you're in a game whether you like it or not.

Using the language of evolution, success in the survival game equates to "fitness." The fittest survive and procreate. Obviously some individuals score better in this game than others. Biologists like to describe such differences in fitness in geographic terms, using the metaphor of a landscape. Using this metaphor, you can think of fitness—or the goal of a game—as getting a good vantage point, living on the peak of a mountain with a good view of your surroundings. For convenience you can describe your fitness just by specifying your latitude and longitude on the landscape map. Some latitude–longitude positions will put you on high ground; some will leave you in a chasm. In other words, some positions are more fit than others. It's just another way of saying that some combinations of features and behaviors improve your chance to survive and reproduce. Real biological fitness is analogous to the better vantage point of a mountain peak.

In a fitness landscape (just like a real landscape) there can, of course, be more than one peak—more than one combination of properties with a high likelihood for having viable offspring. (In the simple landscape of the all-bird island, you'd have a dove peak and a hawk peak.) In a landscape with many fitness peaks, some would be "higher" than others (meaning your odds of reproducing are more favorable), but still many peaks would be good enough for a species to survive.

On a real landscape, your vantage point can be disturbed by many kinds of events. A natural disaster—a hurricane like Katrina, say, or an earthquake and tsunami—can literally reshape the landscape, and a latitude and longitude that previously gave you a great view may now be a muddy rut. Similarly in evolution, a change in the fitness landscape can leave a once successful species in a survival valley. Something like this seems to be what happened to the dinosaurs.

You don't need an asteroid impact to change the biological fitness landscape, though. Simply suppose that some new species moves into the neighborhood. What used to be a good strategy—say, swimming in the lake, away from waterphobic predators—might not be so smart if crocodiles move in. So as evolution proceeds, the fitness landscape changes. Your best evolutionary strategy, in other words, depends on who else is evolving along with you. No species is a Robinson Crusoe alone on an island. And when what you should do depends on what others are doing, game theory is the name of the game.

Recognizing this ever-shifting evolution landscape is the key to explaining how cooperative behavior comes about. In particular, it helps to explain the vastly more elaborate cooperation exhibited by humans compared with other animals.

KIN AND COOPERATION

It's not that nonhuman animals never cooperate. Look at ants, for instance. But such social insect societies can easily be explained by evolution's basis in genetic inheritance. The ants in an ant colony are all closely related. By cooperating they enhance the prospect that their shared genes will be passed along to future colonies.

Similar reasoning should explain some human cooperation— that between relatives. As Maynard Smith's teacher J. B. S. Haldane once remarked, it would make sense to dive into a river to save two drowning siblings or eight drowning cousins. (On average, you share one-half of a sibling's genes, one-eighth of a cousin's.) But human cooperation is not limited to planning family reunion picnics. Somehow, humans evolved to cooperate with strangers.

When I visited Martin Nowak, he emphasized that such nonkin cooperation was one of the defining differences between humans and the rest of the planet's species. The other was language. "I think humans are really distinct from animals in two different ways," he said. "One is that they have a language which allows us to talk about everything. No other animal species has evolved such a system of unlimited communication. Animals can talk about a lot of things and signal about a lot of things to each other, but it seems that they are limited to a certain finite number of things that they can actually tell each other."

Humans, though, have a "combinatorial" language, a mix-and-match system of sounds that can describe any number of circumstances, even those never previously encountered. "There must have been a transition in evolution," Nowak said, that allowed humans to develop this "infinite" communication system. Such a flexible language system no doubt helped humans evolve their other distinction—widespread cooperation. "Humans are the only species that have solved the problem of large-scale cooperation between nonrelated individuals," Nowak pointed out. "That cooperation is interesting because evolution is based on competition, and if you want survival of the fittest, this competition makes it difficult to explain cooperation."[13]

Charles Darwin himself noted this "altruism" problem. Behaving altruistically—helping someone else out, at a cost to you with no benefit in return—does seem to be a rather foolish strategy in the struggle to survive. But humans (many of them, at least) possess a compelling instinct to be helpful. There must have been some survival advantage to being a nice guy, no matter what Leo Durocher might have thought. (He was the baseball manager of the mid-20th century who was famous for saying "Nice guys finish last.")

One early guess was that altruism works to the altruist's advantage in some way, like mutual backscratching. If you help out your neighbor, maybe someday your neighbor will return the favor. (This is the notion of "reciprocal altruism.") But that explanation doesn't take you very far. It only works if you will encounter the recipient of your help again in the future. Yet people often help others whom they will probably never see again.

Maybe you can still get an advantage from being nice in an indirect way. Suppose you help out a stranger whom you never see again, but that stranger—overwhelmed by your kindness—becomes a traveling Good Samaritan, rendering aid to all sorts of disadvantaged souls. Someday maybe one of the Samaritan's beneficiaries will encounter you and help you out, thanks to the lesson learned from the Samaritan you initially inspired.

Such "indirect reciprocity," Nowak told me, had been mentioned long ago by the biologist Richard Alexander but was generally dismissed by evolutionary biologists. And on the face of it, it sounds a little far-fetched. Nowak, though, had explored the idea of indirect reciprocity in detail with the mathematician Karl Sigmund in Vienna. They had recently published a paper showing how indirect reciprocity might actually work, using the mathematics of game theory (in the form of the Prisoner's Dilemma) to make the point. The secret to altruism, Nowak suggested, is the power of reputation. "By helping someone we can increase our reputation," he said, "and to have a higher reputation in the group increases the chance that someone will help you."

The importance of reputation explains why human language became important—so people could gossip. Gossip spreads reputation, making altruistic behavior based on reputation more likely. "It's interesting how much time humans spend talking about other people, as though they were constantly evaluating the reputations of other people," Nowak said. "Language helped the evolution of cooperation and vice versa. A cooperative population makes language more important. . . . With indirect reciprocity you can either observe the person, you can look at how he behaves, or more efficiently you can just talk to people. . . . Language is essential for this."[14]

Reputation breeds cooperation because it permits players in the game of life to better predict the actions of others. In the Prisoner's Dilemma game, for instance, both players come out ahead if they cooperate. But if you suspect your opponent won't cooperate, you're better off defecting. In a one-shot game against an unknown opponent, the smart play is to defect. If, however, your opponent has a well-known reputation as a cooperator, it's a better idea to cooperate also, so both of you are better off. In situations where the game is played repeatedly, cooperation offers the added benefit of enhancing *your* reputation.

TIT FOR TAT

Gossip about reputations may not be enough to create a cooperative society, though. Working out the math to prove that indirect reciprocity can infuse a large society with altruistic behavior turned up some problems. Nowak and Sigmund's model of indirect reciprocity was criticized by several other experts who pointed out that it was unlikely to work except in very small groups. When I next encountered Nowak, in 2004 at a complexity conference in Boston, his story had grown more elaborate.

In his talk, Nowak recounted the role of the Prisoner's Dilemma game in analyzing evolutionary cooperation. The essential backdrop was a famous game theory tournament held in 1980, organized by the political scientist Robert Axelrod at the University of Michigan. Axelrod conceived the brilliant idea of testing the skill of game theoreticians themselves in a Prisoner's Dilemma contest. He invited game theory experts to submit a strategy for playing Prisoner's Dilemma (in the form of a computer program) and then let the programs battle it out in a round-robin competition. Each program played repeated games against each of the other programs to determine which strategy would be the most "fit" in the Darwinian sense.

Of the 14 strategies submitted, the winner was the simplest—an imitative approach called tit for tat, submitted by the game theorist Anatol Rapoport.[15] In a tit-for-tat strategy, a player begins by cooperating in the first round of the game. After that, the player does whatever its opponent did in the preceding round. If the other player cooperates, the tit-for-tat player does also. Whenever the opponent defects, though, the tit-for-tat player defects on the next play and continues to defect until the opponent cooperates again.

In any given series of games against a particular opponent, tit for tat is likely to lose. But in a large number of rounds versus many different opposition strategies, tit for tat outperforms the others on average. Or at least it did in Axelrod's tournament.

Once tit for tat emerged as the winner, it seemed possible that even better strategies might be developed. So Axelrod held a second tournament, this time attracting 62 entries. Of the contestants in the second tournament, only one entered tit for tat. It was Rapoport, and he won again.

You can see how playing tit for tat enhances opportunities for cooperation in a society. A reputation as a tit-for-tat player will induce opponents to cooperate with you, knowing that if they do, you will. And if they don't, you won't.

Alas, the story gets even more complicated. Just because tit for tat won Axelrod's tournament, that doesn't mean it's the best strategy in the real world. For one thing, it rarely won in head-to-head competition against any other strategy; it just did the best overall (because strategies that defeated tit for tat often lost badly against other strategies).

In his talk at the conference, Nowak explored some of the nuances of the tit-for-tat strategy in a broader context. At first glance, tit for tat's success seems to defy the Nash equilibrium implication that everyone's best strategy is to always defect. The mathematics of evolutionary game theory, based on analyzing an infinitely large population, seems to confirm that expectation. However, Nowak pointed out, for a more realistic finite population, you can show that a tit-for-tat strategy, under certain circumstances, can successfully invade the all-defect population.

But if you keep calculating what would happen if the game continues, it gets still more complicated. Tit for tat is an unforgiving strategy—if your opponent meant to cooperate but accidentally defected, you would then start defecting and cooperation would diminish. If you work out what would happen in such a game, the tit-for-tat strategy becomes less successful than a modified strategy called "generous tit for tat." So a generous tit-for-tat strategy would take over the population.

"Generous tit for tat is a strategy that starts with cooperation, and I cooperate whenever you cooperate, but sometimes I will cooperate even when you defect," Nowak explained. "This allows me to correct for mistakes—if it's an accidental mistake, you can correct for it."[16]

As the games go on, the situation gets even more surprising, Nowak said. The generous tit-for-tat approach gets replaced by a strategy of full-scale cooperation! "Because if everybody plays generous tit for tat, or tit for tat, then nobody deliberately tries to defect; everybody is a cooperator." Oh Happy Days.

Except that "always cooperate" is not a stable strategy. As soon as everybody cooperates, an always-defect strategy can invade, just like a hawk among the doves, and clean up. So you start with all defect, go to tit for tat, then generous tit for tat, then all cooperate, then all defect. "And this," said Nowak, "is the theory of war and peace in human history."[17]

GAMES AND PUNISHMENT

Nevertheless, humans do cooperate. If indirect reciprocity isn't responsible for that cooperation, what is? Lately, one popular view seems to be that cooperation thrives because it is enforced by the threat of punishment. And game theory shows how that can work.

Among the advocates of this view are the economists Samuel Bowles and Herbert Gintis and the anthropologist Robert Boyd. They call this idea "strong reciprocity." A strong reciprocator rewards cooperators but punishes defectors. In this case, a more complicated game illustrates the interaction. Rather than playing the Prisoner's Dilemma game—a series of one-on-one encounters—strong reciprocity researchers conduct experiments with various versions of public goods games.

These are just the sorts of games, described in Chapter 3, that show how different individuals adopt different strategies—some are selfish, some are cooperators, some are reciprocators. In a typical public goods game, players are given "points" at the outset (redeemable for real money later). In each round, players may contribute some of their points to a community fund and keep the rest. Then each player receives a fraction of the community fund. A greedy player will donate nothing, assuring a maximum personal payoff, although the group as a whole would then be worse off. Altruistic players will share some of their points to increase the payoff to the whole group. Reciprocators base their contributions on what others are contributing, thereby punishing the "free riders" who would donate little but reap the benefits of the group (but in so doing punish the rest of the group, including themselves, as well). As we've seen, humankind comprises all three sorts of players. Further studies suggest why the human race might have evolved to include punishers.

In one such test of a public goods game,[18] most players began by giving up an average of half their points. After several rounds, though, contributions dropped off. In one test, nearly three-fourths of the players donated nothing by round 10. It appeared to the researchers that people became angry at others who donated too little at the beginning, and retaliated by lowering their own donations—punishing everybody. That is to say, more of the players became reciprocators.

But in another version of the game, a researcher announced each player's contribution after every round and solicited comments from the rest of the group. When low-amount donors were ridiculed, the cheapskates coughed up more generous contributions in later rounds. When nobody criticized the low donors, later contributions dropped. Shame, apparently, induced improved behavior.

Other experiments consistently show that noncooperators risk punishment. So it may have been in the evolutionary past that groups containing punishers—and thus more incentive for cooperation—outsurvived groups that did not practice punishment. The tendency to punish may therefore have become ingrained in surviving human populations, even though the punishers do so at a cost to themselves. ("Ingrained" might not be just in the genes, though—many experts believe that culture transmits the punishment attitude down through the generations.)

Of course, it's not so obvious what form that punishment might have taken back in the human evolutionary past. Bowles and Gintis have suggested that the punishment might have consisted of ostracism, making the cost to the punisher relatively low but still inflicting a significant cost on the noncooperator. They show how game theory interactions would naturally lead societies to develop with some proportion of all three types—noncooperators (free riders), cooperators, and punishers (reciprocators)—just as other computer simulations have shown. The human race plays a mixed strategy.

Still, experts argue about these issues. I came across one paper showing that, in fact, altruism could evolve solely through benefits to the altruistic individual, not necessarily to the group, based on simulations of yet another popular game. Known as the ultimatum game, it is widely used today in another realm of game theory research, the "behavioral game theory" explored by scientists like Colin Camerer. Behavioral game theorists believe that getting to the roots of human social behavior—understanding the Code of Nature—ultimately requires knowing what makes individuals tick. In other words, you need to get inside people's heads. And the popular way of doing that has spawned a hybrid discipline uniting game theory, economics, psychology, and neuroscience in a controversial new discipline called neuroeconomics.

5

Freud's Dream

Games and the brain

The intention is to furnish a psychology that shall be a natural science: that is, to represent psychical processes as quantitatively determinate states of specifiable material particles, thus making those processes perspicuous and free from contradiction.

—Sigmund Freud, Project for a Scientific Psychology, 1895

Sigmund Freud really wanted to understand the brain.

He studied medicine and specialized in neurology. He planned to decipher the code linking the brain's physical processes to the mysteries of the mind. In 1895, he outlined a project for "a scientific psychology," in which mental states and human behavior could be explained materialistically, in terms of the physical interaction of nerve cells in the brain. But Freud found the brain science of the late 19th century too immature to link cranial chemistry to thought and behavior. So he skipped the brain and went straight to the mind, analyzing dreams for clues to the unconscious memories that manipulate mental life.

Others never even dreamed of achieving the "brain physics" that Freud envisioned. Many simply regarded the brain as off limits, declaring it to be a "black box" inaccessible to scientific scrutiny. These "behaviorists" decreed that psychology should stick to observing behavior, studying stimulus and response.

As the 20th century progressed, both Freudianism and behaviorism faded. The black box concealing the brain turned translucent as molecular medicine revealed some of its inner workings. Nowadays the brain is almost transparent, thanks to a variety of scanning technologies that produce images of the brain in action. And so the infant neuroscience that Freud abandoned over a century ago has now matured, nearly to the point of fulfilling his original intention.

Freud could not have dreamed about merging neuroscience with economics, though, for he died before the rise of game theory. And even though they regarded game theory as a window into human behavior, game theory's originators themselves did not imagine that their math would someday advance the cause of brain science. The original game theorists would not have predicted that game theory could someday partner with neuroscience, or that such a partnership would facilitate game theory's quest to conquer economics.[1] But in the late 1990s, game theory turned out to be just the right math for bringing neuroscience and economics together, in a new hybrid field known as neuroeconomics.

BRAINS AND ECONOMICS

One of the appealing features of game theory is the way it reflects so many aspects of real life. To win a game, or survive in the jungle, or succeed in business, you need to know how to play your cards. You have to be clever about choosing whether to draw or stand pat, bet or pass, or possibly bid nillo. You have to know when to hold 'em and know when to fold 'em. And usually you have to think fast. Winners excel at making smart snap judgments. In the jungle, you don't have time to calculate, using game theory or otherwise, the relative merits of fighting or fleeing, hiding or seeking.

Animals know this. They constantly face many competing choices from a long list of possible behaviors, as neuroscientists Gregory Berns and Read Montague have observed (in language rather more colloquial than what you usually find in a neuroscience journal). "Do I chase this new prey or do I continue nibbling on my last kill?" Berns and Montague wrote in *Neuron*. "Do I run from the possible predator that I see in the bushes or the one that I hear? Do I chase that potential mate or do I wait around for something better?"[2]

Presumably, animals don't deliberate such decisions consciously, at least not for very long. Hesitation is bad for their health. And even if animals could think complexly and had time to do so, there's no obvious way for them to compare all their needs for food, safety, and sex. Yet somehow animal brains add up all the factors and compute a course of action that enhances the odds of survival. And humans differ little from other animals in that regard. Brains have evolved a way to compare and choose among behaviors, apparently using some "common currency" for valuing one choice over others. In other words, not only do people have money on the brain, they have the neural equivalent of money operating within the brain. Just as money replaced the barter system— providing a common currency for comparing various goods and services—nerve cell circuitry evolved to translate diverse behavioral choices into the common currency of brain chemistry.

When you think about it, it makes a lot of sense. But neuroscientists began to figure it all out only when they joined forces with economists inspired by game theory. Game theory, after all, was the key to quantifying the fuzzy notion of economic utility. Von Neumann and Morgenstern showed how utility could be rigorously defined and derived logically from simple axioms, but still thought of utility in terms of money. Economists continued to consider people to be "rational" actors who would make behavioral choices that maximized their money or the monetary value of their purchases.

Putting game theory into experimental action, though, showed that people don't always do that. Money—gasp—turned out not to be everything, after all. And people turned out not to be utterly rational, but pretty darn emotional. Imagine that.

GAMES AND EMOTIONS

You might think (and some people do) that game theory therefore becomes irrelevant to the real world of human social interaction, because people are not rational seekers of maximum utility, as game theory allegedly assumes. But while game theory is often described in that way, it's not quite the right picture. Game theory actually only tells you what people *would* do if they *were* "rationally" maximizing their utility. That makes game theory the ideal instrument for identifying deviations from that notion of rationality, and many game theorists are happy with that.

There is, however, another interpretation of what's going on. Perhaps people really do maximize their utility—but utility is not really based on dollars and cents, at least not exclusively. And maybe "emotional" and "rational" are not mutually exclusive descriptions of human behavior. Is it really so irrational to behave in a way that makes you feel good, even if it costs you money? After all, the root notion of utility was really based on happiness, which is surely an emotional notion.

Actually, most economists have long recognized that people are emotional. But when your goal is describing economics scientifically—and mathematically—acknowledging emotions poses a real problem, as Colin Camerer explained to me. "One of the things mainstream economists have said is, well, rationality is mathematically precise," he said. "There's one way to be rational. But there are a lot of ways to not be rational. So they've often used that as an excuse—anything can happen if people aren't perfectly rational." And if anything can happen, there's no hope of finding a mathematical handle on the situation. "Economists have been a little defeatist about this—if you give up rationality, we'll never be able to have anything precise."

This argument seems very much like the strategy of looking for lost keys only under the lamp post, because you couldn't see them if they were anywhere else. If there's only one sort of behavior (rational) that you can describe with your math, then that's the behavior you will assume is correct. But Camerer and other behavioral economists would rather first figure out what behavior is actually like. "Our view is to say, let's find scientists who have been thinking about how brains actually work . . . and ask them for some help," Camerer said. "It might be that even though, mathematically, there are lots of possible alternative models, the psychologists say, 'oh, it's this one.'"[3]

Of course, there was a time—as in Freud's day—when psychologists couldn't have provided very reliable answers to the questions about brain processes underlying human behavior. But with the rise of modern neuroscience, that situation has changed. Human emotions, for instance, are no longer as much of a mystery as they used to be. Scientists can now peer inside the brain to observe what's going on when people experience contempt and disgust, fear or anger, empathy and love. Not to mention getting high on drugs. The driving forces of human decision making can now be traced to signals traveling between specific brain regions. Consequently human behavior, economic and otherwise, can now be analyzed in terms other than the economist's "rational" and monetary notion of utility. In fact, it now seems likely that the brain measures utility not with dollars, but with dopamine. And that's just one of the insights that the new discipline of neuroeconomics is providing into human economic behavior.

ECONOMICS AND THE BRAIN

I had encountered a few papers on neuroeconomics, but really didn't get the big picture until 2003, when I visited Read Montague's laboratory, at the Baylor College of Medicine in Houston. His "Human Neuro-imaging Laboratory" is a cutting-edge model of advanced technology in the service of science, with 100 or so computers, walls lined with plasma screen monitors, and state-of-the-art brain scanning machines. Montague explains it all with the speed of a Pentium processor, emphasizing the power of this new science to grasp human behavior in a precise way.

"We're quantifying the mind and human experience," he said. "We're turning feelings into numbers."[4]

Montague began his scientific life in mathematics and biophysics, but foresight warned him that physics was not the wave of the future. While dabbling in a quantum chemistry project, his thoughts turned to the brain. Why not put math to use in comprehending cognition as well as the cosmos? He began to work on computational modeling of brain processes, and then proceeded to peer deep into real brains, exploiting a technology provided by physics to revolutionize psychology.

Brain scanners are so familiar today that it's hard to remember that a generation or so ago many scientists still considered the brain to be forever inscrutable. The behaviorist psychology of the early 20th century, proselytized by B. F. Skinner, had left its imprint on general beliefs about brain and behavior. Brains could not be observed in action, so only the behaviors that the brain produced mattered to science, the behaviorists contended. It turned out to be a misguided notion of both science and the brain.

By the 1970s, imaginative new technologies had begun to make the brain transparent to clever neurovoyeurs. Radioactive atoms could be attached to critical molecules, allowing their activity to be observed in living brains, providing clues to what brains were doing while animals were behaving. Later methods dispensed with the radioactivity, using magnetic fields to jostle molecules in the blood that flowed through brain tissue. Ultimately this method, known as magnetic resonance imaging, or MRI, became widely used in medicine to "see" beneath the skin. And a variant of MRI technology was adopted by researchers in neuroscience to watch brains in action.[5]

"It can make a movie of the dynamic blood flow changes in every region of your brain," Montague said. And blood flow has been shown to be tightly linked to neural activity—active neurons need nourishment, so that's where the blood goes. You can watch how patterns of activity change in different parts of the brain as its owner performs various behaviors.

Consequently, the old limits on which aspects of the brain could be studied and understood had dissolved, Montague explained, as a new wave of neuroscientists embraced the imaging tools. "There's a kind of sea change of belief in what you can and can't explain," he said. "People put people into scanners like this and do every manner of cognitive task, literally from having sex to thinking about the word *sailboat*. The experiments are working beautifully. I think the sky's the limit."[6]

A new scientific discipline to exploit these technological abilities seems to have emerged almost out of nowhere. The term *neuroeconomics* itself apparently first appeared in 2002.[7] Before that, people like Montague had been referring to their studies as "neural economics." In any event, the first attention-getting published paper in the new genre appeared in 1999, reporting a study by Paul Glimcher and Michael Platt of the Center for Neural Science at New York University. Glimcher and Platt had measured nerve-cell activity in the brains of monkeys performing a decision-making task. The results supported the notion that nervous activity reflects choice-making factors—that is, something like utility—that economists had already identified.

Monkeys, of course, are not obsessed with money, but they do really enjoy getting squirts of fruit juice and can be fairly easily trained to perform all sorts of tasks for a juice-squirt reward. In the Platt-Glimcher experiment, all a monkey was required to do was switch its gaze from a cross on a screen to one of two lights. Looking at a light earned a squirt of juice.

Looking at one of the lights, though, earned a bigger squirt than looking at the other. It didn't take the monkey long to figure that out. (If I'm going to maximize my utility, the monkey obviously thought, I should look at the light on the right.) If the experimenters changed the high-reward squirt to the other light, the monkey caught on right away and preferred the new high-reward light.

None of that was very surprising—similar experiments had been done before. But in this case, Platt and Glimcher also recorded the activity of a nerve cell in a region of the monkey brain that processes visual input and is involved in directing eye movement. (If you must know, the cell was in the lateral intraparietal cortex, or LIP.)

Now here's the tricky part of the experiment. The lights on the screen were positioned so that only one of them was in the field of view accessible to the nerve cell being monitored. When the accessible light appeared, that nerve cell fired electrical impulses, as nerve cells do when stimulated. That nerve cell also boosted its activity as the monkey's eyes moved to gaze at that light. No surprises there. But if that light happened to be the "high reward" light, the nerve cell fired its signals much more vigorously than when viewing the "low reward" light. To an old-school neurophysiologist, that *would* be surprising. For the actual visual stimulus was precisely the same in either case—a light comes on, and the eyes move to look at it. Somehow the neuron linked to that visual stimulus "knew" which light was the Big Gulp of juice dispensers. The monkey's choice of looking toward the high-reward light (that is, the utility-maximizing choice) reflected a specific change in activity by a nerve cell in a specific region of the brain.[8]

Of course, that experiment was just a start, but it opened a lot of scientists' eyes to the possibility of understanding economic decision making by looking inside the brain. The next year, neuroeconomics pioneers met in Princeton for the first major conference on the topic. Montague recalls the skepticism expressed by one of the economists attending, who saw no reason to believe that brain chemicals had anything to do with economics. "I said that is just complete poppycock," Montague recalled. "If your brain doesn't generate economic behavior, what kind of ghost horses do you believe in?" Even worse, the economist didn't even think his remarks were particularly provocative. "I was stunned by that," said Montague. "I might still be stunned by that."[9]

Gradually, though, the idea of merging neuroscience and economics caught on, though perhaps more rapidly in neuroscience than economics. A special issue of *Neuron*, published in October 2002, included a passel of papers on human decision making, many of them exploring the new insights offered by neural economic studies.

Montague and Berns's paper in that issue argued that the

chemical dopamine was the brain's currency for gauging the relative payoffs of potential behaviors. The paper noted various lines of evidence supporting the idea that a circuit of activity linking two parts of the brain—one at the front, behind the forehead, and another deep in the brain's middle—helps govern choice making by producing more or less dopamine. Dopamine levels predict the likely reward associated with different choices, the evidence indicated.

Dopamine had long been known as the brain's chief pleasure molecule, linked to behavior that produces pleasant feelings. But it's not merely pleasure that drives dopamine production. Actually, the brain's dopamine currency seems tuned to the *expectation* of pleasure (or reward of some sort). Some of the brain's dopamine-producing nerve cells are programmed to monitor the difference between expected and actual reward, Montague and Berns showed. If a choice produces precisely the predicted reward, the dopamine cells maintain a constant level of activity. When pleasure exceeds expectations, the cells squirt out dopamine like crazy. If the reward disappoints, dopamine production is curtailed. This monitoring system also takes timing into account—if dinner is delayed, dopamine is diminished. When the anticipated rewards aren't realized, the dopamine monitoring system tells the brain to change its behavior. In this way the expectation of reward can guide a brain's decisions.

A critical point, noted by Montague and Berns, is that all brains are not alike. One person's dream reward might be another's horrific nightmare. Some people make a risky choice only when expecting a huge reward; others gamble for the fun of it. Part of the promise of neuroeconomics is its ability to identify such individual differences with brain scanning.

In one experiment described by Montague and Berns, people chose either A or B on a computer screen and then watched a bar on the screen to see whether their choice earned a reward. (The bar recorded accumulated reward "points" as the game progressed.) As the game went on, the computer adjusted the rewards, based on the player's choices. At first, choosing A raised the bar more, but choosing A too often made B a better bet. When A's payoffs dropped, some players noticed right away and quickly switched to choosing B more often. But others stuck with A, gambling that it would return to its previous high-payoff rate. It appeared that some brains are more inclined to take risks than others—some players play conservatively; others are risk-takers. (Actually, Montague said, more accurate labels for the two types of players would be "matchers" and "optimizers." "I call them conservative and risky because you can make good jokes about that," he said.)

To me, it sounds more like they should be called "switchers" and "stickers." But the labels don't really matter. The most intriguing result from this experiment is the revelations from the brain scans. Sure enough, patterns of brain activity differed in the two groups, particularly in a small clump of brain cells called the nucleus accumbens. It's a brain region implicated in drug addiction, and it's more active in the "risk-taking" game players (the stickers).

The neatest thing, though, is that you can tell who the risk takers and play-it-safers are from their brain scans just after the very beginning of the game, *even while their behaviors are still identical.* This is the sort of evidence that destroys the old behaviorist position that behavior is the only thing that matters (or that you can know). Early in the game, two players can behave identically, making exactly the same choices. Yet by looking into their brains you can see differences that allow you to predict how they will play later, when the payoff rate changes.

"The people that ended up on average being risky are different from these people right away—nobody even jumps categories," Montague told me. Even more intriguing, there appears to be a genetic difference between the two groups as well.

So neuroeconomics thus offers economists a tool they had not possessed before, giving hope that by getting inside people's heads, science might really be on the road to finding the Code of Nature that governs human behavior.

WHOM DO YOU TRUST?

An important advance along that road came in 2003 with the publication of a paper in the journal *Science* by researchers at Princeton
University. In a study by Alan Sanfey and colleagues, participants
in an experiment played the ultimatum game, one of the favorites
of behavioral game theorists. It's kind of like a TV game show
contest in which you are given a lot of money, but you have to
share your windfall with a stranger. Suppose you get $100. You
then offer the stranger part of the money and keep the rest—
unless the stranger refuses your offer. Then you have to give all the
money back, and nobody wins anything.

In theory, the stranger should take any offer, no matter how
small, in order to get something rather than nothing. Therefore, a
game theorist might conclude, you should offer a low amount—
$10, say, or even $1—so that you will then walk away with the
most money possible. But in practice, most strangers reject low
offers. If you offer $10, for instance, you're much more likely to
walk away with zero than $90, as the stranger will probably reject
your offer just to punish you, even at personal expense. Consequently people typically share more generously—offering 40 to
50 percent of the prize, say—in anticipation of an angry rejection
of an unfair offer.

So this is another case where naive game theory, in assuming
that everybody will maximize their money, makes an incorrect prediction, as many economic experiments had already established.
The Princeton study went further, though, by scanning the brains
of the strangers who were considering whether to accept the offer
from the other participant. In this case, the prize was only $10—
science doesn't have budgets like *Who Wants to Be a Millionaire?*—
but the principle was the same. If the first player offered only $1
or $2, the offer was usually rejected. But not always. And you
could tell who was likely to accept or reject a low offer by watching what went on inside their brains.

Stronger brain activity in the front part of a brain region
known as the insula (an area known to be associated with negative

emotions, such as anger and disgust) was common in players who were more likely to reject low offers. Another brain structure—the anterior cingulate cortex—also showed increased activity in those who rejected unfair offers. That region is known to be involved in monitoring conflict—in this case, the conflict between the choice of punishing a cheapskate or turning away money. "Unfair treatment . . . can lead people to sacrifice sometimes considerable financial gain in order to punish their partner for the slight," Sanfey and his collaborators reported in *Science*.[10]

In a commentary on that paper, Colin Camerer noted that it showed how the tenets of basic game theory do not always hold— people do not always act totally in their own self-interest (that is, maximizing their money), and all the players in a "game" therefore are not always trying to do the best they can do, as assumed in the underlying basis for a Nash equilibrium. But behavioral game theory, Camerer noted, can relax these assumptions and still learn a lot about human behavior. The neuroeconomics enterprise, in other words, is a powerful tool for developing behavioral game theory insights into how real people make choices.

Montague's subjects at Baylor, for instance, play similar behavioral games that reveal the quirks of human economic behavior. In one such game—a task for testing trust—Player 1 is given $20 and is allowed to keep some of it and put the rest in a virtual pot, where the amount is then tripled. If Player 1 keeps $10 and donates $10, the sum in the pot becomes $30. Player 2 then gets to split the pot with Player 1—or take it all.

"If you split it 15-15, then in a sense you've repaid the trust," said Montague. But if you take $29 and leave $1, Player 1 is not likely to offer much in the next round of the game. At any point in the game, one player or the other could decide to keep all the money, so the logical move is to take it all as soon as possible, before the other player does. But in fact, players typically trust each other not to be so selfish—although some are more trusting, and some more selfish, than others.

Traditional economists were not surprised at the results of such games. In the 1980s, game theory had fueled the rise of "experimental economics" in which such deviations from pure self-interest showed up regularly. What's new in neuroeconomics is eavesdropping on the players' brains via the MRI scanners while the games are in progress. Montague's lab is particularly well equipped for this sort of thing, with a pair of scanners, one each in two rooms separated by the scientists' observing station. The scientists watch as computers record the brain activity of players deciding how to move or how to react to another player's move. "You can see what went on in the behavior. You can back up and look at their intent to act badly or their intent to invest more," Montague said. "It allows us to cross-correlate what's going on in the two brains. I think it's cool. I think it's an obvious way to study social interactions."[11]

Neuroeconomics does not always require scanning, though. Paul Zak, director of the Center for Neuroeconomics Studies at Claremont Graduate University in California, sometimes uses blood tests instead of brain scans. He can relate variant economic behaviors to levels of certain hormones. In one of Zak's versions of the trust game, players communicate via computer. One player, given $10, offers some of it to another player, who is paid triple the amount offered. (So if Player 1 offers $5, Player 2 gets $15). Player 2 then can take it all, or give part of it back to Player 1. But in this version of the experiment, the game ends after just one round. There's no incentive to earn trust so as to get more money the next time around.

So standard game theory suggests that Player 2 would take all the money, having nothing to gain by giving some back. But Player 1, anticipating that move, should therefore offer none of the money to begin with. Nevertheless, many players defy naive game theory and show at least some trust that the other player will play fair. About half of the first-movers offer some money (suggesting that they are trusting souls), while three in four of the responders give some back (suggesting that they are trustworthy).

Once again, the intriguing thing about such games is finding out what's behind the differences in individual behavior. It turns out that among the trustworthy players, blood tests revealed higher levels of oxytocin, a hormone linked to pleasure and happiness. Apparently the trusting gesture of the first player, by offering some money, elicits a positive hormonal response. "It tells us that people are very much responsive to their environment," Zak told me when I visited him at Claremont. "People who got a positive signal had a nice positive happy hormone response, and their behavior reflects that."[12]

Zak believes that the relationship between trust and oxytocin is central to understanding many of the world's economic ills. Oxytocin is linked to happiness, and the countries where people report high levels of happiness are also countries where people report high degrees of trust. Trust levels, in turn, are a good indicator of a country's economic well-being. "Trust is among the biggest things economists have ever found that are related to economic growth," Zak said.

HOMO NEUROECONOMICUS

For all of its intriguing findings, neuroeconomics doesn't excite everybody, like the economist who perplexed Montague by not caring about the brain. From the perspective of economists like that one, neuroeconomics probably doesn't have much to offer. To them, it only matters what people do; it doesn't matter which part of the brain is busy when they do it.

Neuoreconomists, though, want more than a mere description of economic decision making. They want the Code of Nature, the scientific understanding of humanity sought by 18th-century thinkers such as David Hume and Adam Smith. "The more ambitious aim of neuroeconomics," writes neuroeconomist Aldo Rustichini, "is going to be the attempt to complete the research program that the early classics (in particular Hume and Smith) set out in the first place: to provide a unified theory of human behavior."[13]

Rustichini, of the University of Minnesota, points out that Adam Smith's great works—*Theory of Moral Sentiments* and *Wealth of Nations*—were part of a grand plan to codify the nature of human civilization, to explain how selfish individuals manage to cooperate sufficiently well to establish elaborate functioning societies. Smith's basic answer was the existence of sympathy—the ability of one human to understand what another is feeling. Modern neuroscience has begun to show how sympathy works, by identifying "mirror neurons," nerve cells in the brain that fire their signals both in performing an action and when viewing someone else performing that same action.

Other neuroscientific studies have identified the neural basis of both individual behavioral propensities and collective and cooperative human behavior. Scientists scanning the brains of players participating in a repeated Prisoner's Dilemma game, for instance, have identified regions in the brain that are active in players who prefer cooperating rather than the "purely rational" choice to defect.[14]

Another study used a version of the trust game to examine the brains of people who punish those who play uncooperatively (by keeping all the money instead of returning a fair share). In this game, players who feel cheated may assess a fine on the defector (even though they must pay the price of reducing their own earnings by half the amount of the fine they impose). People who choose to fine the defector display extra activity in a brain region associated with the expectation of reward. That suggests that some people derive pleasure from punishing wrongdoers—the payoff is in personal satisfaction, not in money. In the early evolution of human society, such "punishers" would serve a useful purpose to the group by helping to ostracize the untrustworthy noncooperators, making life easier for the cooperators. (Since this punishment is costly to the individual but beneficial to the group as a whole, it is known as "altruistic punishment.")[15]

Such studies highlight an essential aspect of human behavior that a universal Code of Nature must accommodate—namely that people do not all behave alike. Some players prefer to cooperate while others choose to defect, and some players show a stronger desire than others to inflict punishment. A Code of Nature must accommodate a mixture of individually different behavioral tendencies. The human race plays a mixed strategy in the game of life. People are not molecules, all alike and behaving differently only because of random interactions. People just differ, dancing to their own personal drummer. The merger of economic game theory with neuroscience promises more precise understanding of those individual differences and how they contribute to the totality of human social interactions. It's understanding those differences, Camerer says, that will make such a break with old schools of economic thought.

"A lot of economic theory uses what is called the representative agent model," Camerer told me. In an economy with millions of people, everybody is clearly not going to be completely alike in behavior. Maybe 10 percent will be of some type, 14 percent another type, 6 percent something else. A real mix.

"It's often really hard, mathematically, to add all that up," he said. "It's much easier to say that there's one kind of person and there's a million of them. And you can add things up rather easily." So for the sake of computational simplicity, economists would operate as though the world was populated by millions of one generic type of person, using assumptions about how that generic person would behave.

"It's not that we don't think people are different—of course they are, but that wasn't the focus of analysis," Camerer said. "It was, well, let's just stick to one type of person. But I think the brain evidence, as well as genetics, is just going to force us to think about individual differences."

And in a way, that is a very natural thing for economists to want to do.

"One of the most central and interesting things in economics is specialization and division of labor," Camerer observed. "And so loosely speaking, the more individual difference there is, the better that might be for the economy—as long as you get people in the right jobs. And so knowing more about individual differences could be very important for areas like labor economics, where one of the central questions is are you matching the right workers to the right jobs."[16]

Zak, who has also performed studies to localize the brain's computing of utility, notes that such work revolutionizes the kinds of questions that economists can study.

"In economics we generally think of this utility function as pretty much uniform across individuals," he said. "Now we can ask all kinds of questions about that. How stable is it, how different is it across people, why do you prefer coffee and I prefer tea? What if the price of coffee went up twice as much, what if you haven't drunk coffee in two weeks? Do you value it more, do you value it less? These are really basic questions that may affect things like how things are priced in the market and it may affect how we design laws."[17]

Yet while neuroeconomics may provide the foundation for understanding individual behavior and differences, it cannot alone provide the Code of Nature, or a science of human behavior like Asimov's psychohistory. History comprises the totality of collective human behavior in various forms of social interaction—politically, economically, and culturally. It's in understanding human culture that science must seek a Code of Nature, and game theory provides the best tool for that task.

6

Seldon's Solution

Game theory, culture, and human nature

Self-interest speaks all sorts of languages and plays all sorts of roles.

—La Rochefoucauld

You don't need to know about game theory to understand the ultimatum game. You just need to be a movie fan.

Decades before economists invented the ultimatum game,[1] something very much like it appeared in the 1941 movie *The Maltese Falcon*. The scene is private detective Sam Spade's apartment. Spade (played by Humphrey Bogart) has just made a deal with the criminal Kasper Gutman (Sydney Greenstreet). Spade will collect $1,000 from Gutman and then presumably will share some of it with Brigid O'Shaughnessy (Mary Astor), the film's femme fatale.

"I'd like to give you a word of advice," Gutman whispers to Spade. "I daresay you're going to give her some money, but if you don't give her as much as she thinks she ought to have, my word of advice is, be careful." Gutman knew that people react negatively to the perception of being treated unfairly. He could have predicted the outcome of ultimatum games without game theory or brain scanners, because he was an astute student of human nature.

So why bother with game theory? If you can figure out human nature just by observing how people behave, whether in the real world or the lab, perhaps game theory is nothing more than superfluous mathematics. Besides, when game theory math incorporates the economists' belief in selfish rationality, it doesn't even predict human behavior correctly.

Actually, though, game theory provides a more sophisticated and quantitative tool for describing human nature than the intuition of criminals. Looked at in the right way, the ultimatum game does not disprove game theory, but expands it. Fairness, trust, and other social conditions do affect how people play games and make economic choices. But that just means that the standard economic notion of self-interest is too restrictive—life is more than money. Game theory's math doesn't really tell you what people want, but rather how people should behave in order to achieve what they want.

As economist Jörgen Weibull observes, reports of game theory's death have been exaggerated. "It has many times been claimed that certain game-theoretic solutions—such as Nash equilibrium . . . —have been violated in laboratory experiments," Weibull writes. "While it may well be true that human subjects do not behave according to these solutions in many situations, few experiments actually provide evidence for this."[2]

Early experiments with tests such as the ultimatum game merely assumed that people wanted to maximize their money—which they often failed to do when playing the game. Such tests do not disprove game theory, though; instead, they suggest that something is wrong with the experimenter's assumptions. Later versions of the ultimatum game attempted to include things like fairness, or, more generally, test how a player's social preferences (that is, concerns for others) influence game decisions. Such factors as altruism and spite, Weibull notes, affect the outcome that players prefer to reach, and they make their choices accordingly.

"Indeed, several laboratory experiments have convincingly—though perhaps not surprisingly for the non-economist—shown that human subjects' preferences are not driven only by the resulting material consequences to the subject."[3] In some cases, social context (say, the norms of a person's peer group) dictates choices that appear inconsistent with both personal self-interest and concern for the welfare of others. "Further analysis of preferences of this type seems highly relevant for our understanding of many social behaviors," Weibull observes.[4]

THE NATURE OF HUMAN NATURE

By getting a grip on the nuances of social preferences, game theory enhances its prospects for forging a science of human behavior, a Code of Nature for predicting social phenomena. But there might be a flaw in that plan. It presumes that there is such a thing as "human nature" to begin with for game theory to describe.

At first glance, experiments such as those using the ultimatum game do seem to provide evidence for a consistent human nature. After all, when economists play the ultimatum game with college students, the results come out pretty much the same, whether in Los Angeles, Pittsburgh, or even Tokyo. And of course, one well-known battalion of social scientists argues strongly that there most definitely is a universal human nature. They are devotees of a discipline known as evolutionary psychology, a widely publicized field contending that human behavior today reflects the genetic selection imposed on the species during the early days of human evolution. Human nature, this notion implies, is a common heritage of the race, shaping the way people instinctively respond to situations today, based on how they behaved in order to survive in hunter-gatherer times.

A typical advocate of this view is Harvard psychologist Steven Pinker, who argued his beliefs with considerable passion in a book called *The Blank Slate*. Viewing the brain as blank at birth, to be shaped totally by experience, is nonsense, he insisted. General features of human nature have been programmed by evolution and stored on a genetic hard drive that guides the brain's development. As a result, human nature today derives from the era of early human evolution. "The study of humans from an evolutionary perspective has shown that many psychological faculties (such as our hunger for fatty food, for social status, and for risky sexual liaisons) are better adapted to the evolutionary demands of our ancestral environment than to the actual demands of the current environment," Pinker wrote.[5]

In other words, people today are just hunter-gatherers wearing suits.

On the surface, it might seem that it would be a good thing for game theory—and the rest of the human sciences—if this idea is right. If the Code of Nature is inscribed into the human genetic endowment, that should improve the prospects for deciphering the rules governing human nature and then predicting human behavior. After all, the concept that a Code of Nature exists might be interpreted to mean that there *is* some universal behavioral program to which all members of the human species conform.

Yet with all due respect to much of the intelligent research that has been done in the field of evolutionary psychology, some of the conclusions that have been drawn from it rest on rather shaky ground. And it turns out that rather than bolstering evolutionary psychology, game theory helps to show why it breaks down. Furthermore, the way game theory does it has much in common with the way that Asimov's fictional hero Hari Seldon found the solution to formulating his physics of society, or psychohistory.

COMPARING CULTURES

In *Prelude to Foundation*, the first prequel to Asimov's Foundation Trilogy, a young Hari Seldon delivers a talk at a mathematics conference on the planet Trantor, capital world of the Galactic Empire. Seldon's talk describes his idea of predicting the future via the math of psychohistory, a science that he had just begun to develop. Naturally the emperor receives word of this talk (in the galactic future, politicians pay more attention to science than they do today) and invited Seldon to an audience.

"What I have done," Seldon told the emperor, "is to show that, in studying human society, it is possible . . . to predict the future, not in full detail, of course, but in broad sweeps; not with certainty, but with calculable probabilities."[6]

But the emperor was dismayed to learn that Seldon couldn't

actually predict the future just yet, that he merely had the germ of an idea about how to do so if the mathematics could be properly developed. Seldon, in fact, was skeptical that he would ever succeed.

"In studying society, we put human beings in the place of subatomic particles, but now there is the added factor of the human mind," Seldon explained. "To take into account the various attitudes and impulses of mind adds so much complexity that there lacks time to take care of all of it."[7]

In fact, Seldon pointed out, an effective psychohistory capable of predicting the galactic future would have to account for the interacting human variables on 25 million planets, each containing more than a billion free-thinking minds. "However theoretically possible a psychohistorical analysis may be, it is not likely that it can be done in any practical sense," he admitted.[8]

By displeasing the emperor with such pessimism, Seldon soon found himself a fugitive, roaming from one sector to another on the planet Trantor—the urban sector of the Imperial capital, a university town, a farming region, an impoverished mining center. By the end of the book Seldon realized that Trantor was a microcosm of the galaxy, home to hundreds of societies each with their own mores and customs. That was his solution to achieving a science of psychohistory! He didn't have to analyze 25 million worlds; he could understand the variations in human behavior by using Trantor itself as a laboratory.

Toward the end of the 20th century, Earth-bound anthropologists independently arrived at a similar scheme for analyzing human social behavior. By playing the ultimatum game (and some variants) in small, isolated societies around the planet, those scientists have found that human nature isn't so universal after all. College students in postindustrial society, it turns out, are not perfectly representative of the entire human race.

This worldwide game-playing project began after anthropologist Joe Henrich, then a graduate student at UCLA, tried out the ultimatum game with the Machiguenga farmers of eastern Peru in 1996. The rules were the same as with college students: One player is given a sum of money and must offer a share of it to the second player. The second player may either accept the offer (and the first player keeps the rest) or the second player may reject the offer, in which case all the money is returned and neither player gets anything.

By the time Henrich tried the game in Peru, it had been widely played with college students, who usually make offers averaging more than 40 percent of the pot. Such offers are routinely accepted. Sometimes lower amounts would be offered, but they would usually be rejected. Among the Machiguenga, though, Henrich observed that lower amounts were routinely offered—and usually accepted.

"We both expected the Machiguenga to do the same as everybody else," UCLA anthropologist Robert Boyd told me. "It was so surprisingly different that I didn't know what to expect anymore."[9]

Could it be that the Machiguenga actually understood the rational-choice rules of game theory, while everybody else in the world let emotions diminish their payoffs? Or would other isolated cultures behave in the same way? Soon Henrich, Boyd, and others acquired funding from the MacArthur Foundation, and later the National Science Foundation, to repeat the games in 15 small-scale societies on four continents. The results were utterly baffling. From Fiji to Kenya, Mongolia to New Guinea, people played the ultimatum game not just the way college students did, or the way economic theory dictated, but any way they darn well pleased.

In some cultures, like the Machiguenga, low offers were typical and were often accepted. But in other cultures, low offers were frequently made but typically rejected. In a few cultures the offers would sometimes be extra generous—even more than half. But in some societies such generous offers were likely to be refused. Among other groups, rejections almost never occurred, regardless of the size of the offer.[10]

"It really makes you rethink the nature of human sociality," Henrich, now at Emory University in Atlanta, told me. "There's a lot of variation in human sociality. Whatever your theory is about human behavior, you have to account for that variation."[11]

CULTURAL DIVERSITY

This cross-cultural game theory research clearly shows that people in many cultures do not play economic games in the selfish way that traditional economic textbooks envision. And it appears that the differences in behavior are indeed rooted in culture-specific aspects of the group's daily life. Individual differences among the members of a group—such as sex, age, education, and even personal wealth—did not affect the likelihood of rejecting an offer very much. Such choices apparently depend not so much on individual idiosyncrasies as on the sorts of economic activity a society engages in. In particular, average offers seemed to reflect a society's amount of commerce with other groups. More experience participating in markets, the research suggested, produces not cutthroat competition, but a greater sense of fairness.

The stingy Machiguenga, for instance, are economically detached from most of the world—in fact, they hardly ever interact with anyone outside their own families. So their market-based economic activity is very limited, and their behavior is selfish. In cultures with more "market integration," such as the cattle-trading Orma in Kenya, ultimatum game offers are generally higher, averaging 44 percent of the pot and often are as much as half.

Orma average offers are similar to those found with American college students. But sometimes students make low offers, and the Orma rarely do. College students find their low offers are usually rejected, but in some societies any offer is accepted, no matter how low. Among the Torguud Mongols of western Mongolia, for example, a low offer is rarely refused. Even so, Torguud offers averaged between 30 and 40 percent—despite the fact that the offerer would surely get more by offering less. Apparently the local Mongolian culture values fairness more than money. At the same time, inflicting punishment (by rejecting an offer) is not highly regarded there, either.

In society after society, the anthropologists discovered different ways in which cultural considerations dictated unselfish behavior. Among the Aché of Paraguay, for example, hunters often leave the day's game on the outskirts of their village. Members of the tribe then retrieve it for sharing among the villagers. When playing the ultimatum game, the Aché typically make high offers, often more than half. So do the whale-hunting Lamalera of Indonesia, who carefully and fairly divide up the meat from killed whales.

In other societies, though, the cultural influences play out differently. In Tanzania, the Hadza share meat, but they complain about it and try to get away without sharing when they can. Nonsharers, though, risk ostracism, social scorn, and negative gossip. It makes sense, then, that when playing the ultimatum game, the Hadza make low offers, with high rejection rates.

On the other hand, high offers do not always signify a culture imbued with altruism. The Au and Gnau of Papua New Guinea often offer more than half the money, but such generosity is frequently rebuffed. The reason, it seems, is that among the Au and Gnau accepting a gift implies an obligation to reciprocate in the future. And an excessively large offer may be interpreted as an insult.

Colin Camerer, one of the economists collaborating with the anthropologists in the cross-cultural games, observes that this result is just another twist in the cultural influence on economic behavior. "Offering too much money, rather than being extremely generous, is actually being kind of mean—it's demeaning," Camerer explained to me. "So the money is turned down because they don't want to be insulted, and they don't want to be in debt."[12]

The surprising results of the cross-cultural game theory experiments showed that the games were not necessarily measuring what the scientists thought they were. Rather than purely testing economic behavior, the games actually tapped into patterns of cultural practice. Players apparently tried to figure out how the game related to their real-world life and then behaved accordingly.

For instance, the Orma quickly recognized a similarity between real life and a variant of the ultimatum experiment, the public goods game (which we encountered in Chapters 3 and 4). In that game the experimenter (Jean Ensminger of Caltech) offered each of four Orma some money from which they could contribute to a community pot and keep the rest. Ensminger would then double the pot and divide it equally among the four players. When she described the game to her Kenyan assistants, they quickly replied that it was just like harambee—a practice of soliciting contributions for community projects.

"That really changed our thinking a lot about what was going on when people are in an experiment," Camerer told me in one of our conversations at Caltech. "In game theory, the bias we inherited was the mathematician's bias." In other words, the initial belief was that "when you present the game, it's like a smart kid sitting down to play Monopoly or poker. . . . They read the rules, figure out what to do—they treat it as like a logic problem. But these subjects treat it as like analogical reasoning—what is this like in my life?"[13]

So what the game theory experiments have shown is that life differs in different cultures, and economic behavior reflects those differences in cultural life. Game theory has consequently illuminated the interplay of culture and economic behavior, showing that humankind does not subscribe to a one-size-fits-all mentality. Human culture is not monolithic—it's like a mixed strategy in game theory.

In an intriguing way, this diversity in cultural behavior around the world parallels the multiplicity of versions of "human nature" found within various academic disciplines. When I visited Boyd in his office—on the third floor of Haines Hall on the UCLA campus—our discussion turned to that problem in pursuing the general notion of human nature and the basic principles of human behavior. Boyd lamented the academic world's fragmented and inconsistent view of how people tick.

"We have this weird, I think untenable, situation in the social sciences," he said. "You go over to Bunch Hall and the economists tell the students one thing. And the students come over here to sociology, one floor down, and they get told no, that's all wrong, this is right. And they come up here, and we anthropologists tell them all kinds of different things. . . . And then they go to the psychology department and they get told something different again. This is not OK. It's not acceptable that the economists are happy with their world and the sociologists are happy with their world, and this persists in an institution which is supposed to be about getting at the truth."[14]

Perhaps the rise of game theory as a social science tool, though, will help change that situation. In particular, merging the abstract math of game theory with the real-world immersion of anthropologists and other social scientists has begun to show how disparate views of human nature may be drawn closer to how life really works.

"Somehow in the last 20 years there's been this emergence," Boyd said, "of people who are interested in doing mathematical theory like game theory, but building it on psychologically real people."

GAMES, GENES, AND HUMAN NATURE

The fairness displayed in many societies and the variety of behaviors among them are hard to reconcile with the view that human psychology is universally programmed by the evolutionary past. A hard-line interpretation of evolutionary psychology would predict similar behavior everywhere. The game experiment project argues otherwise, posing a conundrum for evolutionary psychologists.

"I think that if it had turned out that everywhere in the world people were . . . ruthlessly selfish, they would have said, 'See, I told you so,'" said Boyd. "And when it didn't turn out that way . . . that's not a comfortable fact for them. It's some fairly strong evidence on the other side of the scale." He pointed out, though, that evolution remains important to human psychology. "No educated person should doubt that our psychology is the product of evolution— that's a given," Boyd said. "The question is, how did it work?"

And as Camerer pointed out, evolutionary psychologists can always retreat to the fallback position that the ancestral environment programmed people to be different. But in that case the original claim about a single "human nature" is substantially softened. "I think the hard story about cultural universality, you can reject," Camerer said.[15]

It's important to perceive, I think, that these are not the knee-jerk reactions against "genetic determinism" expressed by some enemies of evolutionary psychology and its intellectual predecessor, sociobiology. These are evidence-driven conclusions about evolutionary psychology's limitations. While evolutionary psychology has benefited from a surge of often favorable publicity over the last decade or so, more and more thoughtful critiques (as opposed to vitriolic polemics) have begun to appear.

One of the more interesting critiques comes from philosopher David Buller, of Northern Illinois University in Dekalb, who critically assessed the methodological rigor underlying several of evolutionary psychology's claimed "successes" and found that the evidence for them was actually ambiguous. In a book published in 2005 and in a paper published the same year in *Trends in Cognitive Sciences*, Buller distinguished the mere study of evolution's relationship to psychology—evolutionary psychology with a lowercase e and p—from Evolutionary Psychology, the paradigm based on the "doctrine of a universal human nature" and the "assumption that the adaptational architecture of the mind is massively modular."

"Evolutionary Psychologists argue that our psychological adaptations are 'modules,' or special-purpose 'minicomputers,' each of which evolved during the Pleistocene to solve a problem of survival or reproduction faced by our hunter-gatherer ancestors," Buller wrote.[16]

He contends that many of the "discoveries" claimed by evolutionary psychologists crumble under critical analysis. Evolutionary psychologists say their work explains sex differences in jealousy, an innate ability to detect "cheating" (as when someone fails to perform an obligation incurred in return for receiving some benefit), and a tendency of parents to abuse stepchildren more than their own genetic offspring. But however plausible the Evolutionary Psychology explanations might be, Buller says, the actual evidence underlying them suffers from a number of defects. In some cases the data on which the claims are based may be biased or incomplete, and sometimes the research methods are not rigorous enough to exclude alternative explanations for the findings. Buller argues, for example, how results of a card-choosing task, designed to illustrate the brain's "cheating detector" module, could also be explained by a nonmodular brain just acting logically. "Although the Evolutionary Psychology paradigm is a bold and innovative explanatory framework, I believe it has failed to provide an accurate understanding of human psychology from an evolutionary perspective," he wrote.[17]

Buller's criticisms reflect the latest stage of a long-running controversy about the role of genes and evolution in shaping human culture and patterns of behavior, an issue commonly framed as a battle of nature versus nurture—genes versus environment. The Evolutionary Psychology view ascribes enormous power to the role of genetic endowment in directing human behavior; many scientists, philosophers, and scholars of other stripes find the belief in the dictatorial determinism of genetic power to be particularly distasteful.

In any case, objections such as Buller's—whether they turn out to be well founded or not—should not be regarded as support for the extreme view (sometimes still expressed, surprisingly) that rejects any role for genes in behavior—or more precisely, in differences among humans in their behavior. Without genes, of course, there is no behavior—because there would be no brain, and no body, to begin with. The real question is whether variations in individual genetic makeup contribute to the wide variety of behavioral tendencies found among people and cultures. In recent years, the most thoughtful investigators of this issue have tended to agree that genes do matter, to some degree or another. Anyone who says that genes don't matter at all has clearly not been paying attention to modern molecular genetics research, particularly in neuroscience. And modern neuroscience does even provide some evidence for modularity in many brain functions, as Evolutionary Psychologists argue. But the latest neuroscience also undercuts the Evolutionary Psychology paradigm in a major way by showing how flexible the brain is. A brain hardwired for certain behaviors ought to be, in fact, hardwired. But the human brain actually exhibits remarkable flexibility (the technical term is plasticity) for adapting its tendencies in the wake of experience.

"One of the surprises of the last few years is the fact that we're learning that the brain is hardwired for change," says Ira Black, of the Robert Wood Johnson Medical School in New Jersey. "We've learned that the environment is capable of accessing genes and altering their activity within the brain."[18]

Heredity does wire some predispositions into the brain, to be sure, but it's a mistake to believe that experience must somehow defy the brain's genetic hardwiring. It is actually the brain's genetic wiring that creates the capacity to change with experience. "You are flexible because of your genes, not in spite of them," declare neuroscientists Terrence Sejnowski and Steven Quartz in their book *Liars, Lovers, and Heroes*. "Your experiences with the world alter your brain's structure, chemistry, and genetic expression, often profoundly, throughout your life."[19]

So most experts would agree that genes are important, and genetic variation can influence propensities toward different kinds of behavior. On the other hand, genes are not so all-powerfully important as some gene-power dogmatists contend. Even animals, often portrayed as mere "gene machines" responding to stimuli with programmed responses, actually exhibit a lot of variability in their behavior that cannot be ascribed to genetic variations.

A few years back I ran across a study that put this issue in particularly sharp perspective, having to do with an especially simple behavioral response in mice. For years, scientists have annoyed mice by dipping their tails into a cup of hot water (typically about 120 degrees Fahrenheit). The idea is to test a mouse's reaction to pain. Sure enough, the mice do not like having their tails dipped into hot water; as soon as you put the tail in, the mouse will jerk it out.

But not all mice behave in exactly the same way—at least, not all pull their tails out as rapidly as others. Experimenters have found that some mice react, on average, in a second or less; others might take three or four. Some mice are simply more sensitive to pain than others. Since the environmental conditions are apparently just the same, it is tempting to conclude that differences in this simple behavior reflect some difference in the mice's genes. It's an easy enough question to check: Since the experiments are performed on different genetic strains of mice, all you need to do is compare the results for the different strains to see if some genetic profiles corresponded with slower (or faster) tail-jerk reactions than others.

As it turns out, Jeffrey Mogil of McGill University in Montreal and collaborators at the University of Illinois had been dipping mouse tails in hot water for more than a decade and had accumulated plenty of data with which to perform such an analysis. And that analysis did confirm the relevance of genetic differences. Keep the environmental conditions constant (the water temperature should be precisely 49 degrees Celsius, for example) and some genetic strains, on average, do flip their tails out of the water faster than others.

Upon further review, though, it became clear that genes were not the only things that mattered, and a constant water temperature was not the only environmental factor to consider. After reviewing the scores of more than 8,000 irritated mice, Mogil's team found that all sorts of things influence reaction speed. Are the mice kept in a crowded cage, or do they have room to roam? Was it the first mouse out of the cage, or the second? Is it morning, afternoon, or night? Did anybody remember to measure the humidity? And who was holding the mouse at the time? "A factor even more important than the mouse genotype was the experimenter performing the test," Mogil and colleagues wrote in their paper.[20] In other words, genes aren't even as important as which researcher is handling the mouse.

In fact, a computerized cross-check of all the factors found that genetic differences accounted for only 27 percent of the variation in tail-test reaction speed. Environmental influences were responsible for 42 percent of the performance differences, with 19 percent attributed to interactions between environment and genes.

(That just means that certain conditions influenced some genetic strains but not others.)

Mogil and collaborators concluded that the laboratory environment plays an important role in the way mice behave, either masking or exaggerating the effects under genetic control. And since tail-flipping is such a simple behavior—basically a spinal cord reflex—it's unlikely that the environment's influence in this case is a fluke. More complicated behaviors would probably be even more susceptible to environmental effects, the researchers observed.

Results such as these strike me as similar to findings about how humans play economic games in different ways. Genes, environment, and culture interact to produce a multiplicity of behaviors in mice, and in people. The human race has adopted a mixed strategy for surviving in the world, with a diverse blend of behavioral types. It shouldn't be surprising that cultures differ around the world as well, that the planet is populated by a "mixed strategy" of cultures, rooted in a mixture of influences on how behavior evolves.

A MIXED HUMAN NATURE

So what of human nature, and game theory's ability to describe it? There is a human nature, but it is not the simplistic consistent human nature described by extreme evolutionary psychologists. It is the mixed human nature that, on reflection, should be obvious in a world ruled by game theory. Evolution, after all, is game theory's ultimate experiment, where the payoff is survival. As we've seen, evolutionary game theory does not predict that a single behavioral strategy will win the game. That would be like a society populated by all hawks or all doves—an unstable situation, far from Nash equilibrium. Game theory's rules induce instead a multiplicity of strategies, leading to a diverse menagerie of species practicing different sorts of behaviors to survive and reproduce.

Seen through the lens of game theory, evolution's role in human psychology is still important, but it operates more subtly than hard-line evolutionary psychologists have suggested. Game theory guarantees that evolution will produce a diversity of species, a mixture of behaviors, and in the case of the human race, a multiplicity of cultures.

So it seems to me that game theory has itself answered the question about why it doesn't seem to work, at least as it was originally formulated. Nash's original game theory math was construed and interpreted a little too narrowly. Applied solely to economics, it predicted behavior that was often at odds with what people really did. But that was because the math originated and operated in an abstract realm of assumptions and calculations. Now, by playing games around the world with real people enmeshed in their own cultural milieus, scientists have shown how that purely mathematical approach to economics and behavior can be modified by real-world considerations.

"My goal is to get the mathematicians to loosen their grip on game theory and get away from thinking about a game . . . that's purely of mathematical interest," Camerer told me. Instead, he said, playing games can be thought of as something "like an X-ray about a thing that's happening in the world."[21]

Viewed in this way, game theory becomes even more powerful. It becomes a tool for grappling with the complexity of human behavior and understanding the innumerable interactions that drive human history. It's just the sort of thing Hari Seldon was looking for to produce a science of society.

Of course, Asimov's character had many real-life predecessors who sought a similar science of society. In fact, the statistical physics that Asimov cited as the inspiration for psychohistory owed its own inspiration to the pioneers who applied statistics to people—especially an astronomer turned sociologist named Adolphe Quetelet.

7

Quetelet's Statistics and Maxwell's Molecules

Statistics and society, statistics and physics

The mob has many heads but no brains.
—English proverb

The actions of men . . . are in reality never inconsistent, but however capricious they may appear only form part of one vast system of universal order.
—Henry Thomas Buckle

When creating the fictional science of psychohistory, more than half a century ago, Isaac Asimov didn't bother to give the details of how the math worked. He simply said you could describe masses of people in the same way you describe masses of molecules. Trained as a chemist, Asimov knew well that the behavior of gases under different conditions could be calculated with precision, even though nobody could possibly know what any one of that gas's atoms or molecules was doing. And so he reasoned that a sufficiently advanced science could do the same thing with people.

"Psychohistory dealt not with man, but man-masses," Asimov wrote.[1] "It was the science of mobs; mobs in their billions. . . . The reaction of one man could be forecast by no known mathematics; the reaction of a billion is something else again." So while any one person could do his or her own thing, society might collectively exhibit patterns of behavior that equations could capture. Psychohistory might not be quite as accurate as the laws governing gases, but that's only because there are many more molecules than people. As one of Asimov's characters explained, "The laws of history are as absolute as the laws of physics, and if the probabilities of error are greater, it is only because history does not deal with as many humans as physics does atoms, so that individual variations count for more."[2]

Still, psychohistory *was* fiction, and using math to describe something as complex as society still strikes many people as an overly ambitious goal for real life. On the other hand, in the mid-19th century math seemed similarly useless for physicists pondering the complexities of molecular motion in gases. Gross properties of gases could be observed but not understood without a way to quantify the apparent anarchy of molecular interactions. How could anyone grasp the inner workings of a mass of molecules too numerous to count and too small to be seen? Yet the Scottish physicist James Clerk Maxwell found a way, by using statistics—mathematical descriptions of the average behavior of large groups of molecules.

Calculating such averages provided amazing predictive power. Although you couldn't say exactly what any one molecule was up to, you could predict precisely what a sufficiently large group of molecules would do in certain circumstances. Measuring the temperature of a gas, for instance, tells you something about the average speed of its molecules, and you can calculate the effect of altering the temperature on the gas's pressure. Similar methods were developed to deal with matter in all sorts of situations. Knowing the average amount of energy possessed by molecules of various substances, for instance, allows you to predict whether a chemical reaction will proceed or not—and if so, how far. You can use the statistical approach to describe a substance's magnetic or electric properties, or whether it will snap or stretch when under tension. In Asimov's psychohistory, features of society corresponded to variables like the temperature and pressure of a gas or the ebb and flow of chemical reactions or the fracture of a beam in a building.

While Asimov's vision remains a science fiction dream, it is now closer to reality than probably even he would have thought possible. The statistical approach inaugurated by Maxwell has today become physicists' favorite weapon for invading the social sciences and describing human actions with math. Physicists have applied the statistical approach to analyzing the economy, voting behavior, traffic flow, the spread of disease, the transmission of opinions, and the paths people take when fleeing in panic after somebody shouts Fire! in a crowded theater.

But here's the thing. This isn't a new idea, and physicists didn't have it first. In fact, Maxwell, who was the first to devise the statistical description of molecules, got the idea to use statistics in physics from social scientists applying math to society! So before statistical physicists congratulate themselves for showing the way to explaining the social sciences, they should pause to reflect on the history of their field. As the science journalist Philip Ball has observed, "by seeking to uncover the rules of collective human activities, statistical physicists are aiming to return to their roots."[3]

In fact, efforts to apply science and math to society have a rich history, extending back several centuries. And that history contains hints of ideas that can, in retrospect, be seen as similar to key aspects of game theory—foreshadowing an eventual convergence of all these fields in the quest for a Code of Nature.

STATISTICS AND SOCIETY

The idea of finding a science of society long predates Asimov. In a sense it goes back to ancient times, of course, resembling at least partially the old notion of a "natural law" of human behavior or a Code of Nature. In early modern times, the idea received renewed impetus from the success of Newtonian physics, stimulating the efforts of Adam Smith and others as described in Chapter 1. Even before Newton, though, the rise of mechanistic physical science inspired several philosophers to consider a similarly rigorous approach to society.

In medieval times, the importance of the mechanical clock to society conditioned scientists to think of the universe in mechanical terms. Descartes, Galileo, and other pioneers of modern science advocated a mechanical, cause-and-effect view of the cosmos that ultimately led to Newton's definitive system of physics, published in his *Principia* in 1687. It was only natural that the implications of mechanism for life and society attracted the attention of other 17th-century thinkers. One was Thomas Hobbes, whose famous work *Leviathan* described the state of society that (Hobbes believed) maximized the well-being of all its members. Conveniently for Hobbes, a supporter of the British monarchy, his conclusion was that the people should turn over control of society to an absolute monarch. Otherwise, he argued, a dog-eat-dog mentality of unrestrained human nature would guarantee life to be "nasty, brutish, and short."

In an intriguing paper, though (published in *Physica A*), Philip Ball points out that Hobbes's questionable conclusion was not as important as the methods he used to reach it. The Hobbes approach was to assess the interacting preferences of various individuals and figure out how best to achieve the best deal for everybody. The resulting theoretical framework, Ball says, "could be recast without too much effort" as a Nash equilibrium maximizing the power of each individual. As such, Hobbes's *Leviathan* could be seen as an early effort to understand society mathematically, with the prescient indication that something like game theory would be a good mathematical instrument for the task.

Real math entered the story a little later, as the science of statistics was invented—for the very purpose of quantifying various aspects of society. The scientist and politician Sir William Petty, a student of Hobbes, advocated the scientific study of society in a quantitative way. His friend John Graunt began compiling tables of social data, such as mortality figures, in the 1660s. Graunt and others began to keep track of births and deaths and analyzed the data much like the way that baseball fans pore over batting averages today. By a century later, in the times leading up to the French Revolution, gathering social statistics had become a widespread practice, usually undertaken in the belief that studying such social numbers might reveal laws of social nature the same way astronomers had revealed the regularities of the heavens. "The idea that there were laws that stood in relation to society as Newton's mechanics stood in relation to the motion of the planets was shared by many," writes Ball.[4]

Gathering numbers was not enough, of course, to make the study of society a science in the Newtonian mold. Physics, as Newton had sculpted it, was the science of certainty, his dictatorial laws of motion determining how things happened. Statistics dealt not with such certainty, but rather displayed considerable variability. Much about human behavior seemed to depend on chance—the luck of the draw (as in games!). Dealing with people called for quantifying luck—leading to the mathematical analysis of probability.

Early studies of probability theory predated Newton, starting with the mid-17th-century work of Blaise Pascal and Pierre Fermat—their idea being to figure out how to win at dice or card games. An economic use of probability theory soon arose from insurance companies, which used statistical tables to gauge the risk of people dying at certain ages or the likelihood of fires or shipwrecks destroying insured property.

Probability became more useful to physics (and the rest of science) with the development of the theory of measurement errors during the 18th century, particularly in astronomy. Ironically, one of the key investigators in that statistical field was Pierre Simon, Marquis de Laplace, the French mathematician famous for his articulation of Newtonian determinism. For a being with intelligence capable of analyzing the circumstances of all the bodies in the universe, and the forces operating on them, all movements great and small could be foreseen by applying Newton's laws, Laplace declared. "For it, nothing would be uncertain and the future, as the past, would be present to its eyes."[5]

Laplace recognized full well, though, that no human intelligence possessed such grand ability. So statistical methods were needed to deal with the unavoidable uncertainties afflicting human knowledge. Laplace wrote extensively on the issue of probability and uncertainties, focusing especially on the inevitable errors that occurred whenever measurements were made.

Suppose, for instance, that you're trying to measure the positions of the planets visible in the night sky. No matter how good your instruments, uncontrollable factors will prevent you from measuring positions with arbitrary precision. Each time you measure a planet's position, the answer will be at least a little bit off from whatever the true position might be. But such random errors do not render your measurements hopelessly inaccurate. While individual errors might be random, the sum of all errors could be subjected to mathematical analysis in a way that revealed something about the planetary position's true coordinates. If the measurements are careful enough, for example, small errors will be more common than somewhat larger errors, and huge errors would be even rarer.

Laplace was one among several mathematicians who developed the math for calculating the range of such errors. Another was Carl Friedrich Gauss, the German mathematician whose name was given to the now familiar bell-shaped curve that depicts how random measurement errors are distributed around the average value (the "Gaussian distribution").[6] For repeated measurements, the most likely true value would simply be the value at the peak of the curve—the average (or mean) of all the measurements (assuming the "errors" are all due to random, uncontrollable factors, rather than some problem with the instrument itself). The math describing the curve tells you how to estimate the likelihood that the true value differs from the mean by any given amount.

While Gauss got his name on the curve, Laplace's work in this arena turned out to be more important for the human side of statistics. Like others of his era, Laplace recognized the relevance of statistics to human affairs, and applied the error curve to such issues as the ratio of male to female births. Laplace's interest in the social side of his math led to a much broader appreciation of its potential uses—thanks to the Belgian mathematician and astronomer Adolphe Quetelet.

SOCIAL PHYSICS

Quetelet, who was born in Ghent in 1796, made a major mathematical contribution to society that most Americans today are uncomfortably familiar with, although few people know to blame Quetelet. He invented the Quetelet index for assessing obesity, a measure better known now as the Body Mass Index, or BMI. But he had much greater vision for applying science to society than merely telling people how to know when they were overweight.

As a youth, Quetelet dabbled in painting, poetry, and opera, but his special talent was math, and he earned a math doctorate in 1819 at the University of Ghent. He got a job teaching math in Brussels, where he was soon elected to the Belgian academy of sciences. During the 1820s, Quetelet expanded his interest from math into physics, and in 1823 he traveled to Paris to study astronomy, part of a plan to establish an observatory in Brussels.

Quetelet later wrote some widely read popularizations of astronomy and physics for the general reader. And he often delivered public lectures on science attended by all segments of the public. Quetelet was highly regarded as a teacher and as a person by those who knew him—he was described as amiable and considerate, tactful and modest, but still a rigorous thinker who expressed his views strongly.[7]

During his stay in Paris, Quetelet took in more than just astronomy. He also learned probability theory from Laplace and met his colleagues Poisson and Fourier, who also had an interest in the statistics of society. Quetelet was himself strongly attracted to the social sciences, and he soon realized that Laplace's uses of the bell curve to describe social numbers could be dramatically expanded.

Quetelet began to publish papers on the statistical description of society, and in 1835 authored a detailed treatise on what he called social physics[8] (or social mechanics), introducing the idea of an "average man" for analyzing social issues. He knew that there is no one average man, but by averaging various aspects of a great many men, much could be learned about society. "In giving to my work the title of Social Physics, I have had no other aim than to collect, in a uniform order, the phenomena affecting man, nearly as physical science brings together the phenomena appertaining to the material world," Quetelet commented.[9]

Quetelet's key point was that the diversity of behaviors among men, seemingly too complex to comprehend, would coalesce into regular patterns when assessed for vast numbers. "In a given state of society, resting under the influence of certain causes, regular effects are produced, which oscillate, as it were, around a fixed mean point, without undergoing any sensible alterations," he wrote.[10] The same statistical laws governing measurement errors could be enlisted to disclose predictable patterns underlying the apparent chaos of historical trends and events, he believed.

Understanding the "average man," Quetelet contended, was essential for sound government based on an intelligent understanding of human nature. No single set of attributes regarded as the defining features of human nature would apply in all respects to any given individual, of course. Yet certain tendencies would show up in society more often than others, so statistical methods could be used to construct an abstract "average" representation of the typical mix of human characteristics.

Quetelet illustrated his point with the analogy of an archery target. After many shots by many archers, the arrows lodged in the target form a distinct pattern, some close to the bull's-eye, some farther away. But suppose for some reason that the outline defining the bull's-eye was obscured. Even if no arrow had actually hit it, you could infer the bull's-eye's location from the pattern made by the arrows. "If they be sufficiently numerous, one may learn from them the real position of the point they surround," Quetelet pointed out.[11]

Quetelet collected all the data he could find on such social variables as birth and death rates, analyzing how such rates differed by location, season, and even time of day. He cataloged and assessed evidence of the influences of moral, political, and religious factors on crime rates. He was struck by the constancy of crime reports in various sorts of categories from year to year. In any given locale, for instance, the number of murders remained pretty much the same from one year to the next, and even the murderer's methods showed a similar distribution.

"The actions which society stamps as crimes," Quetelet wrote, "are reproduced every year, in almost exactly the same numbers; examined more closely, they are found to divide themselves into almost exactly the same categories; and if their numbers were sufficiently large, we might carry farther our distinctions and subdivisions, and should always find there the same regularity."[12] Similarly, the rate of crimes for different ages displayed a constant distribution, with the 21–25 age group always topping the list. "Crime pursues its path with even more constancy than death," Quetelet observed.[13]

He warned, though, of the dangers posed by interpreting such statistics without sufficiently careful thought. Another researcher, for instance, had shown that property crime in France was higher in provinces where more children were sent to schools, and concluded that education caused crime. It's the sort of reasoning you hear today on talk radio. Quetelet correctly chastised such stupidity.

Quetelet also repeatedly emphasized that the statistical approach could not be used to draw conclusions about any given individual (another obvious principle that is often forgotten by today's media philosophers). The insurance company's mortality tables cannot forecast the time of any one person's death, for instance. Nor can any single case, however odd, invalidate the general conclusions drawn from a statistical regularity.

Quetelet's exposition of social statistics attracted a great deal of attention among scientists and philosophers. Many of them were aghast that he seemed to have little regard for the supposed free will that humans exercised as they pleased. Quetelet responded not by denying free will, but by observing that it had its limits, and that human choice was always influenced by conditions and circumstances, including laws and moral strictures. In making the simplest of choices, Quetelet noted, our habits, needs, relationships, and a hundred other factors buffet us from all sides. This "empire of causes" typically overwhelms free will, which is why, with knowledge of all the factors affecting someone's choice, it is usually possible to predict what it will be.

In any event, the controversy over Quetelet's views served science well, for it guaranteed that his work was to become widely known. Fortunately for physics, some of the commentaries on it reached the hands of James Clerk Maxwell.

MAXWELL AND MOLECULES

Maxwell was one of those once-in-a-century geniuses who perceived the physical world with sharper senses than those around him. He saw deeply into almost every corner of physics, forever alert to the hidden principles governing the complexities of physical phenomena. He mastered electricity and magnetism, light and heat, pretty much mopping up all the major areas of physics beyond those that Newton had already taken care of—gravitation and the laws of motion. And in fact, Maxwell detected an essential shortcoming in Newton's laws of motion, too. They worked fine for macroscopic objects, like cannonballs and rocks. But what about the submicroscopic molecules from which such objects were made? Presumably, Newton's laws would still apply. But they did you no good, because you could not possibly trace the motion of an individual molecule anyway. And if you couldn't describe the motion of an object's parts, how could you expect to predict the behavior of the object?

For a cannonball dropped from the leaning tower of Pisa, the internal motion of the metal's atoms made no difference to the rate at which it fell. But other forms of matter did not cooperate so willingly. Suppose you wanted to know how changes in pressure affected the temperature of the steam in a steam engine, for instance. You could not even begin to calculate the motions of the individual H_2O molecules.

Physicists were not helpless in the face of this question—they had devised some pretty good formulas for describing how gases behaved. Maxwell, though, wanted to know how those rules worked—why gases behaved the way they did. If he could show

how molecular motions produced the observed gross behavior, he would have achieved both a deeper understanding of the phenomena *and* have provided strong new evidence for the very existence of atoms and molecules, which was at the time—in the mid-19th century—a contentious conviction in some circles.

The idea that a gas's properties depended on the motion of its constituent molecules was not new, though. It was known as the kinetic theory of gases, originally articulated in 1738 by our old friend Daniel Bernoulli, who explained the gas laws with a crude picture of molecules modeled as billiard balls. But as the science historian Stephen Brush has noted, Bernoulli's theory "was a century ahead of its time."[14] Bernoulli's idea was based on the (correct) notion that heat is merely the motion of molecules, but in his day most physicists believed heat to be some sort of fluid substance (called caloric). By the 1850s, though, the kinetic theory was a ripe topic for physicists to study, as the laws of thermodynamics—constituting the correct theory of heat—were arriving on the scene.

One of the major pioneers of thermodynamics was the German physicist Rudolf Clausius. In an 1857 paper, Clausius presented a comprehensive view on the nature of heat as molecular motion. He described how the pressure of a gas was related to the motion of molecules as they impinged on the walls of their container. And any given molecule was constantly battered by collisions with other molecules, so its behavior reflected the influences of such impacts (just as a person's choices reflect the influences of the countless social pressures that Quetelet had described). In his approach, Clausius emphasized the importance of the average velocity of the molecules, and in an 1858 paper introduced the important notion of the average distance that molecules traveled between collisions (a distance labeled the "mean free path").

In 1859, Maxwell entered the molecular motion arena, exploring the interplay of gas molecules and their resulting velocities a little more deeply. In his approach, Maxwell applied the sort of statistical thinking that Quetelet had promoted.

Maxwell had probably first encountered Quetelet in an article by the astronomer John Herschel. (Herschel, of course, was familiar with Quetelet as a fellow astronomer.) Later, in 1857, Maxwell read a newly published book by the historian Henry Thomas Buckle. Buckle, himself clearly influenced by Quetelet, believed that science could discover the "laws of the human mind" and that human actions are part of "one vast system of universal order."[15] (I encountered one Web page where Buckle is referred to as the Hari Seldon of the 19th century.)

Buckle was another of the 19th century's most curious characters. Born near London in 1821, he was a slow learner as a child. When he was 18 his father, a maritime merchant, died, leaving the son sufficient funds to tour Europe and pursue his hobbies of history and chess. (Buckle became a formidable chess player and learned several foreign languages, becoming fluent in seven and conversant in a dozen others. He also became a prolific bibliophile, amassing a library exceeding 20,000 books.)

From 1842 on, Buckle began compiling the data and evidence for a comprehensive treatise on history. Originally planned to focus on the Middle Ages, the work eventually took on broader aims and became the *History of Civilization in England* (by which Buckle actually meant the history of civilization, period). While presumably a work of history, Buckle's book was really more a sociological attempt to subject the nature of human behavior to the methods of science. He criticized the "metaphysical" (or philosopher's) approach to the issue, advocating instead the "historical" method (by which he basically meant the scientific method).

"The metaphysical method . . . is, in its origin, always the same, and consists in each observer studying the operations of his own mind," Buckle wrote. "This is the direct opposite of the historical method; the metaphysician studying one mind, the historian studying many minds."[16] Buckle could not resist remarking that the metaphysical method "is one by which no discovery has ever been made in any branch of knowledge." He then emphasized the need for observing great numbers of cases so as to escape the effects of "disturbances" obscuring the underlying law. "Every thing we at present know," Buckle asserted, "has been ascertained by studying phenomena, from which all causal disturbances having been removed, the law remains as a conspicuous residue. And this can only be done by observations so numerous as to eliminate the disturbances."[17]

Much of Buckle's philosophy echoes Quetelet, including similar slams against the idea of unfettered free will. Occasionally someone makes what appears to be a free and even surprising choice, but only because you don't know enough about the person's circumstances, Buckle observed. "If, however, I were capable of correct reasoning, and if, at the same time, I had a complete knowledge both of his disposition and of all the events by which he was surrounded, I should be able to foresee the line of conduct which, in consequence of those events, he would adopt," Buckle pointed out.[18] Read retrospectively, Buckle's comment sounds very much like what a game theorist would say today. Game theory is, in fact, all about understanding what choice would (or should) be made if all the relevant information influencing the outcome of the decision is known.

Buckle realized that choices emerge not merely from external factors, though, but from the inner workings of the mind as well. Since sorting out the nuances of all the influences exceeds science's powers, the nature of human behavior must be described instead by the mathematics of statistics. "All the changes of which history is full, all the vicissitudes of the human race, their progress or their decay, their happiness or their misery, must be the fruit of a double action; an action of external phenomena upon the mind, and another action of the mind upon the phenomena," wrote Buckle. "The most comprehensive inferences respecting the actions of men are derived from this or from analogous sources: they rest on statistical evidence, and are expressed in mathematical language."[19]

It's not hard to imagine Maxwell reading these words and seeing in them a solution to the complexities confounding the description of gases. Though Maxwell found Buckle's book "bumptious," he recognized it as a source of original ideas, and the statistical reasoning that Buckle applied to society seemed just the thing that Maxwell needed to deal with molecular motion. "The smallest portion of matter which we can subject to experiment consists of millions of molecules," Maxwell later noted. "We cannot, therefore, ascertain the actual motion of any one of these molecules; so that we are obliged to . . . adopt the statistical method of dealing with large groups of molecules."[20] That statistical method, he showed, could indeed reveal "uniformities" in molecular behavior. "Those uniformities which we observe in our experiments with quantities of matter containing millions of millions of molecules are uniformities of the same kind as those explained by Laplace and wondered at by Buckle," Maxwell declared.[21]

The essential feature of Maxwell's work was showing that the properties of gases made sense not if gas molecules all flew around at a similar "average" velocity, as Clausius had surmised, but only if they moved at all sorts of speeds, most near the average, but some substantially faster or slower, and a few very fast or slow. As the molecules bounced off one another, some gained velocity; others slowed down. In subsequent collisions, a fast molecule might be either slowed down or speeded up. A few would enjoy consecutive runs of very good (or very bad) luck and end up moving extremely rapidly (or slowly), while most would get a mix of bounces and tend toward the overall average velocity of all the molecules in the box.

Just as Quetelet's average man was fictitious, and key insights into society came from analyzing the spread of features around the average, understanding gases meant figuring out the range and distribution of molecular velocities around the average. And that distribution, Maxwell calculated, matched the bell-shaped curve describing the range of measurement errors.

As Maxwell refined his ideas during the 1860s, he showed that when the velocities reached the bell-shaped distribution, no further net change was likely. (The Austrian physicist Ludwig Boltzmann further elaborated on and strengthened Maxwell's results.) Any specific molecule might speed up or slow down, but the odds were strong that other molecules would change in speed to compensate. Thus the overall range and distribution of velocities would stay the same. When a gas reached that state—in which further collisions would cause no net change in its overall condition—the gas was at equilibrium.

Of course, this notion of equilibrium is precisely analogous to the Nash equilibrium in game theory. And it's an analogy that has more than merely lexical significance. In a Nash equilibrium, the sets of strategies used by the participants in a game attain a stable set of payoffs, with no incentive for any player to change strategies. And just as the Nash equilibrium is typically a mixed set of strategies, a gas seeks an equilibrium state with a mixed distribution of molecular velocities.

PROBABILITY DISTRIBUTIONS

Nash's mixed strategies, and Maxwell's mixed-up molecules, are both examples of what mathematicians call probability distributions. It's such an important concept for game theory (and for science generally) that it's worth a brief interlude to mercilessly pound the idea into your brain (possibly with a silver hammer). Consider Maxwell's problem. How do the molecules in a gas share the total amount of energy that the gas possesses? One possibility is that all the molecules move at something close to the average, as Clausius suspected. Or the velocities could be distributed broadly, some molecules leisurely floating about, others zipping around at superspeed. Clearly, there are lots of possible combinations. And all of these allocations of molecular velocities are in principle possible. It's just that some combinations of velocities are more likely than others.

For a simpler example, imagine what happens when you repeatedly flip a coin 10 times and record the number of heads. It's easy to calculate the probability distribution for pennies, because you know that the odds of heads versus tails are 50-50. (More technically, the probability of heads for any toss is 0.5, or one-half. That's because there are two possibilities—equally likely, and the sum of all the probabilities must equal 1—1 signifying 100 percent of the cases.) In the long run, therefore, you'll find that the average number of heads per trial is something close to 5 (if you're using a fair coin). But there are many conceivable combinations of totals that would give that average. Half the trials could turn up 10 heads, for instance, while the other half turned up zero every time. Or you could imagine getting precisely 5 heads in every 10-flip trial.

What actually happens is that the number of trials with different numbers of heads is distributed all across the board, but with differing probabilities—about 25 percent of the time you'll get 5 heads, 20 percent of the time 4 (same for 6), 12 percent of the time 3 (also for 7). You would expect to get 1 head 1 percent of the time (and no heads at all out of a 10-flip run about 0.1 percent of the time, or once in a thousand). Coin tossing, in other words, produces a probability distribution of outcomes, not merely some average outcome. Maxwell's insight was that the same kind of probability distribution governs the possible allocations of energy among a mess of molecules. And game theory's triumph was in showing that a probability distribution of pure strategies—a mixed strategy—is usually the way to maximize your payoff (or minimize your losses) when your opponents are playing wisely (which means they, too, are using mixed strategies).

Imagine you are repeatedly playing a simple game like matching pennies, in which you guess whether your opponent's penny shows heads or tails. Your best mixed strategy is to choose heads half the time (and tails half the time), but it's not good enough just to average out at 50-50. Your choices need to be made randomly, so that they will reflect the proper probability distribution for equally likely alternatives. If you merely alternate the choice of heads or tails, your opponent will soon see a pattern and exploit it; your 50-50 split of the two choices does you no good. If you are choosing with true randomness, 1 percent of the time you'll choose heads 9 times out of 10, for instance.

In his book on behavioral game theory, Colin Camerer discusses studies of this principle in a real game—tennis—where a similar 50-50 choice arises: whether to serve to your opponent's right or left side. To keep your opponent guessing, you should serve one way or the other at random.[22] Amateur players tend to alternate serve directions too often, and consequently do not achieve the proper probability distribution. Professionals, on the other hand, do approach the ideal distribution more nearly, suggesting that game theory does indeed capture something about optimal behavior, and that humans do have the capability of learning how to play games with game-theoretic rationality.

And that, in turn, makes a point that I think is relevant to the prospects of game theory as a mathematical method of quantifying human behavior. In many situations, over time, people do learn how to play games in a way so that the results coincide with Nash equilibrium. There are lots of nuances and complications to cope with, but at least there's hope.

STATISTICS RETURNS TO SOCIETY

Of course, real-life situations, the rise of civilization, and the evolution of culture and society are much more complicated than flipping coins and playing tennis. But that is also true of the inanimate world. In most realms of physics and chemistry, the phenomena in need of explanation are rarely split between two equally likely outcomes, so computing probability distributions is much more complicated than the simple 50-50 version you can use with pennies. Maxwell, and then Boltzmann, and then the American physicist J. Willard Gibbs consequently expended enormous intellectual effort in devising the more elaborate formulas that today are known as statistical mechanics, or sometimes simply statistical physics. The uses of statistical mechanics extend far beyond gases, encompassing all the various states of matter and its behavior in all possible circumstances, describing electric and magnetic interactions, chemical reactions, phase transitions (such as melting, boiling, freezing), and all other manner of exchanges of matter and energy.

The success of statistical mechanics in physics has driven the belief among many physicists that it could be applied with similar success to society. Nowadays, using statistical physics to study human social interactions has become a favorite pastime of a whole cadre of scientists seeking new worlds for physics to conquer. Everything from the flow of funds in the stock market to the flow of traffic on interstate highways has been the subject of statistical-physics study.

So the use of statistical physics to describe society is not an entirely new endeavor. But the closing years of the 20th century saw an explosion of new research in that arena, and as the 21st century opened, that trend turned into a tidal wave. Behind it all was a surprising burst of new insight into the mathematics describing complex networks. The use of statistical physics to describe such networks has propelled an obscure branch of math called "graph theory" into the forefront of social physics research. And it has all come about because of a game, starring an actor named Kevin Bacon.

8

Bacon's Links

Networks, society, and games

Unlike the physics of subatomic particles or the large-scale structure of the universe, the science of networks is the science of the real world—the world of people, friendships, rumors, disease, fads, firms, and financial crises.

—Duncan Watts, *Six Degrees*

Modern science owes a lot to a guy named Bacon.

If you had said so four centuries ago, you would have meant Francis Bacon, the English philosopher who stressed the importance of the experimental method for investigating nature. Bacon's influence was so substantial that modern science's birth is sometimes referred to as the Baconian revolution.

Nowadays, though, when you mention Bacon and science in the same breath you're probably talking not about Francis, but Kevin, the Hollywood actor. Some observers might even say that a second Baconian revolution is now in progress.

After all, everybody has heard by now that Kevin Bacon is the most connected actor in the movie business. He has been in so many films that you can link almost any two actors via the network of movies that he has appeared in. John Belushi and Demi Moore, for instance, are linked via Bacon through his roles in *Animal House* (with Belushi) and *A Few Good Men* (with Moore). Actors who never appeared with Bacon can be linked indirectly: Penelope Cruz has no common films with Bacon, but she was in *Vanilla Sky* with Tom Cruise, who appeared with Bacon in *A Few Good Men*. By mid-2005, Bacon had appeared in films with nearly 2,000 other actors, and he could be linked in six steps or fewer to more than 99.9 percent of all the linked actors in a database dating back to 1892. Bacon's notoriety in this regard has become legendary, even earning him a starring role in a TV commercial shown during the Super Bowl.

Bacon's fame inspired the renaissance of a branch of mathematics known as graph theory—in common parlance, the math of networks. Bacon's role in the network of actors motivated mathematicians to discover new properties about all sorts of networks that could be described with the tools of statistical physics. In particular, modern Baconian science has turned the attention of statistical physicists to social networks, providing a new mode of attack on the problem of forecasting collective human behavior.

In fact, the new network math has begun to resemble a blueprint for a science of human social interaction, a Code of Nature. So far, though, the statistical physics approach to quantifying social networks has mostly paid little attention to game theory. Many researchers believe, however, that there is—or will be—a connection. For game theory is not merely the math for analyzing individual behavior, as you'll recall—it also proscribes the rules by which many complex networks form. What started out as a game about Kevin Bacon's network may end up as a convergence of the science of networks and game theory.

SIX DEGREES

In the early 1990s, Kevin Bacon's ubiquity in popular films caught the attention of some college students in Pennsylvania. They devised a party game in which players tried to find the shortest path of movies linking Bacon to some other actor. When a TV talk show publicized the game in 1994, some clever computer science students at the University of Virginia were watching. They soon launched a research project that spawned a Web page providing instant calculations of how closely Bacon was linked to any other actor. (You should try it—go to *oracleofbacon.org*.) The 1,952 actors directly linked by a common film appearance with Bacon each have a "Bacon number" of 1. Another 169,274 can be linked to Bacon through one intermediary, giving them a Bacon number of 2. More than 470,000 actors have a Bacon number of 3. On average, Bacon can be linked to the 770,269 linkable actors in the movie database[1] in about 2.95 steps. And out of those 770,269 in the database, 770,187 (almost 99.99 percent) are linked to Bacon in six steps or fewer—nearly all, in other words, are less than six degrees of separation from Bacon.

So studies of the Kevin Bacon game seemed to verify an old sociological finding from the 1960s, when social psychologist Stanley Milgram conducted a famous mail experiment. Some people in Nebraska were instructed to send a parcel to someone they knew personally who in turn could forward it to another acquaintance with the eventual goal of reaching a Boston-area stockbroker. On average, it took a little more than five mailings to reach the stockbroker, suggesting the notion that any two people could be connected, via acquaintances, by less than "six degrees of separation." That idea received considerable publicity in the early 1990s from a play (and later a movie) of that title by John Guare.

From a scientific standpoint, the Bacon game and Guare's play came along at a propitious time for the study of networks. The six-degrees notion generated an awareness that networks could be interesting things to study, just when the tools for studying networks fell into scientists' laps, in the form of powerful computers that, it just so happened, were themselves linked into a network of planetary proportions—the Internet.

NETWORKS ARE US

When I was growing up, "network" meant NBC, ABC, or CBS. Later came PBS, CNN, and ESPN, among others, but the basic idea stayed the same. As the world's cultural focus shifted from TV to computers, though, the notion of network expanded far beyond its origins. Nowadays it seems that networks are everywhere, and everything is a network. Networks permeate government, the environment, and the economy. Society depends on energy networks, communication networks, and transportation networks. Businesses engage networks of buyers and sellers, producers and consumers, and even networks of insider traders. You can find networks of good ol' boys—in politics, industry, and academia. Atlases depict networks of rivers and roads. Food chains have become food *webs*, just another word for networks. Bodies contain networks of organs, blood vessels, muscles, and nerves. Networks are us.

Of all these networks, though, one stands out from the crowd—the Internet and the World Wide Web. (OK, that's actually two networks; the Internet comprising the physical network of computers and routers, while the World Wide Web is technically the software part, consisting of information on "pages" connected by URL hyperlinks.) During the early 1990s, awareness of the Internet and Web spread rapidly through the population, bringing nearly everybody in contact with a real live example of a network in action. People in various walks of life began thinking about their world in network terms. True, the word "network" already had its informal uses, for such things as groups of friends or business associates. But during the closing years of the 20th century, the notion of network became more precise and came to be applied to all sorts of systems of interest in biology, technology, and society.

Throughout the scientific world, networks inspired a new viewpoint for assessing some of society's most perplexing problems. Understanding how networks grow and evolve, survive or fail, may help prevent e-mail crashes, improve cell phone coverage, and even provide clues to curing cancer. Discovering the laws governing networks could provide critical clues for how to protect—or attack—everything from power grids and ecosystems to Web sites and terrorist organizations. Physicists specializing in network math have infiltrated disciplines studying computer systems, international trade, protein chemistry, airline routes, and the spread of disease.

Using math to study networks is not entirely new, though. In fact, network math goes back at least to the 18th century, when the Swiss mathematician Leonhard Euler gave intellectual birth to the field with his analysis of a network of bridges in Königsberg, in eastern Prussia. In the mid-20th century, Paul Erdös and Alfréd Rényi developed the math to describe networks in their most abstract representation—essentially dots on paper connected by lines. The dots are known as nodes (or sometimes vertices); the lines are officially called edges, but are more popularly referred to as links. Such drawings of dots and lines are technically known as graphs, so traditional network math is known as graph theory.[2]

A graph's dots and edges can represent almost anything in real life. The nodes may be any of various objects or entities, such as people, or companies, or computers, or nations; the links may be wires connecting machines, friendships connecting people, common film appearances connecting movie actors, or any other common property or experience. People, of course, belong to many different kinds of networks, such as networks of family, networks of friends, networks of professional collaborators. There are networks of people who share common investments, political views, or sexual partners.

Traditional graph theory math does not do a very good job of describing such networks, though. Its dots and lines resemble real networks about as much as a scorecard resembles a baseball field. The scorecard does record all the players and their positions, but you won't get much of an idea of what baseball is like from reading the scorecard. Same with graphs. Standard graph math describes fixed networks with nodes connected at random, whereas in the real world, networks usually grow, adding new parts and new connections, while perhaps losing others—and not always at random. In a random network, every node is an equal among many, and few nodes get much more or less than a fair share of links. But in many real-world networks, some nodes possess an unusually high number of links. (In a network of sexual partners, for example, some people have many more "links" than average—an important issue in understanding the spread of HIV.) And real networks form clusters, like cliques of close friends.

Erdös and Rényi knew full well that their dots and lines did not capture the complexities of real-world networks. As mathematicians, they didn't care about reality—they developed their model to help understand the mathematical properties of random connections. Describing random connections was a mathematically feasible thing to do; describing all the complexities of real-world networks was not. Nobody knew how to go about doing it.

But then a paper appearing in the British journal *Nature* began to change all that. Looking back, the birth of network mania can be dated to June 4, 1998, when Duncan Watts and Steven Strogatz published a brief paper (taking up only two and a half *Nature* pages) called "Collective Dynamics of 'Small-World' Networks."[3]

NETWORK MANIA

A few years later, when I met Strogatz at a complexity conference, I asked him why networks had become one of math's hottest topics in the late 1990s. "I think our paper started it," he said. "If you ask me when did this really start, I think it started in 1998 when our paper appeared in *Nature* on what we called small-world networks."

So I quizzed Strogatz about that paper's origins. It really was a case of culture preparing the conditions for the advance of science.

"When Watts and I started our work in 1995 or so, we were very aware of the whole Kevin Bacon thing, and we had heard of six degrees of separation, and the movie had come out of that play," said Strogatz. "So it was in the air."[4]

Of course, Kevin Bacon didn't revolutionize science totally on his own. The Bacon game became famous just about the time that the public became aware of the Internet, thanks to the arrival of the World Wide Web.

"I think the Web got us thinking about networks," Strogatz

said. Not only was the Web a high-profile example of a vast, elaborate network, it made many other networks accessible for research. Web crawlers and search engines made it possible to map out the links of the Web itself, of course, and the Web also made it possible to catalog other large networks and store them for easy access (the database of movie actors being one prime example). Data on the metabolic reactions in nematode worms or the gene interactions in fruit flies could similarly be collected and transmitted.

"Big databases became available, and researchers could get their hands on them," Strogatz observed. "People started to think about things as networks." Before that, he said, even actual networks weren't usually viewed in network terms—the electric power network was known as a grid, and you were just as likely to hear the term telephone "system" as telephone network. "We didn't think of them so much as networks," Strogatz said. "I don't think we had the visceral sensation of moving through a network from link to link."

With the Web it was different. It was almost impossible to think of it as a whole. You had to browse, link by link. And the Web touched all realms of science, linking specialists of all sorts with network ideas. "In many different branches of science," Strogatz observed, "the kind of thinking that we call network thinking started to take hold."

Still, the revolution in network math did not begin until after the Watts-Strogatz paper appeared in 1998. They showed how to make a model of a "small-world" network, in which it takes only a few steps on average to get from any one node of the network to any other. Their model produced some surprises that led to a flurry of media coverage and the subsequent network mania. But Strogatz thinks some of those surprises have been misrepresented as being responsible for network math's revival. Some experts would say, for example, that the Watts-Strogatz paper's major impact stemmed from identifying the small-world nature of some particular real-world networks. Others have suggested that "clustering" of links (small groups of nodes connected more than randomness would suggest) was the key discovery. "This is to me the bogus view of what was important about our paper," Strogatz said. "The reason it caught on, I think, is because we were the first to compare networks from different fields and find that there were similar properties across fields."

In other words, diverse as networks are, many share common features that can be described in a mathematically precise way. Such commonalities gave people hope that network math could be more than a tedious chore of sorting out links in one kind of network and then moving on to the next. Instead, it seemed, general laws of networks might be possible, enabling accurate forecasts of how different kinds of networks would grow, evolve, and behave—chemical networks like proteins in cells, neural networks like nerve cells in brains, or social networks, such as actors in movies or stock traders in the economy.

SMALL WORLDS

One of the key common features of different networks is that many of them do in fact exhibit the small-world property. When a network's nodes are people, for instance, small worlds are the rule. So discovering the rules governing small-world networks may be the key to forecasting the social future.

Watts and Strogatz uncovered the small-world nature of certain networks by focusing on networks intermediate between those that were totally regular or totally random. In a regular network (what is often called a regular lattice), the nodes are connected only to their nearby neighbors. For an ultra-simple example, think of a series of nodes arranged in a circle. The dots representing the nodes are connected to their immediate neighbors on either side by the line representing the circle.

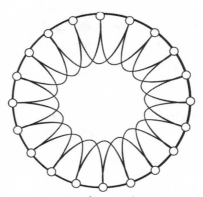

Regular network

For a more elaborate (but still regular) network, you could connect the dots to their second-nearest neighbors as well. Each node would then be connected to four others—two neighbors on each side.

In a random network, on the other hand, some nodes would be connected to many others, some maybe connected to only one. Some nodes would be linked only to other nodes nearby; some would be connected to nodes on the other side of the circle; some would be connected both to neighbors and to distant nodes. It would look like a mess. That's what it means to be random.

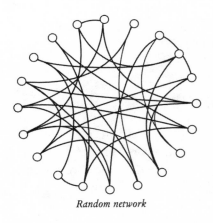

Random network

In a random network, it is usually easy to find a relatively short path from one node to any other, thanks to the random long-range links making connections across the circle. Regular networks, though, are not so easy to navigate. To get from one side of the circle to the other, you have to take the long way around, via linked neighbors.

But what happens, Watts and Strogatz wondered, with an "in-between" network—neither completely regular nor completely random? In other words, suppose you started with a regular network and then added just a few links at random between other nodes. It turned out that if even just a tiny percentage of the links created shortcuts between distant nodes, the new intermediate network would be a small world (that is, you could get anywhere in the network in a small number of steps). But that intermediate network retains an important feature of the regular network—its nearby nodes are still more highly connected than average (that is, they are "clustered"), unlike random networks where clustering is mostly absent.

The mathematical existence of graphs combining these properties of random and regular networks was nice, if not necessarily important. But the fact that you needed very few shortcuts to make the network world small implied that small-world networks might be common in nature. Watts and Strogatz tested that possibility on three real-world examples: the film actors' network starring Kevin Bacon, the electrical power grid in the western United States, and the network of nerve cells in the tiny roundworm *C. elegans*.[5] In all three cases, these networks exhibited the small-world property, just like the models of hypothetical networks that were intermediate between regular and random.

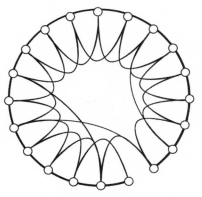

Intermediate (small-world) network

"Thus," Watts and Strogatz concluded, "the small-world phenomenon is not merely a curiosity of social networks nor an artifact of an idealized model—it is probably generic for many large, sparse networks found in nature."[6]

If so (and it was), Watts and Strogatz had opened a new frontier for mathematicians and physicists to explore, where all sorts of important networks could be analyzed with a common set of tools. In just the way that statistical physics made it possible to tame the complexities of a jumble of gas molecules, mathematicians could use similar math to compute a network's defining properties. And just as all gases, no matter what kinds of molecules they contained, obeyed the same gas laws, many networks observed similar mathematical regularities. "Everybody pointed out, isn't this remarkable that these totally different networks have these properties in common—how would you have ever thought that?" Strogatz said.

Several network features can be quantified by numbers analogous to the temperature and pressure of a gas, what scientists call the parameters describing a system. The average number of steps to get from any one node to any other—the "path length"—is one such parameter. Another is the "clustering coefficient"—how likely two nodes are to be directly linked if they are both linked to a third. A relatively high clustering rate is one of the counterintuitive features of small-world networks. The short path length in small-world networks is similar to the situation in random networks. On the other hand, the high clustering coefficient is completely unlike that in random networks, but is more similar to that in regular networks.

This clustering property (you could call it the measure of "cliquishness") is especially of interest in social networks. Since my sister Sue, for instance, has a friend Debby and a friend Janet, it is more likely than average that Debby and Janet also know each other. (They do.) "There is a tendency to form triangles, and you wouldn't see that in random networks," Strogatz pointed out.

Besides clustering coefficient and path length, another critical number is the average number of links connecting one node to another, known as the degree coefficient. (The "degree" of a node is the number of other nodes it is linked to.) As a node in the actor network, Kevin Bacon would be ranked very high in degree, being connected to so many others. Being well connected, after all, is what makes the average path length between Bacon and other actors so short. But in a shocking development, it turned out that Bacon is far from the most connected of actors. Taking the average number of steps to link to another actor as the gauge, he doesn't even rank in the top 1,000!

It turns out, in fact, that Bacon's true importance for networks had nothing to do with how special he is, but rather how typical he is. Many actors, like Bacon, serve as "hubs" connecting lots of other members of the acting community. And the existence of such hubs turns out to be a critical common feature of many real-world networks.

THE POWER OF SCALE FREEDOM

As of mid-2004, the actor leading the list as "most connected" (based on the average number of steps to link him with all the other actors) was Rod Steiger, at 2.679 steps. Bacon, at 2.95, ranked 1,049th. (Still, as Bacon is more connected than 99 percent of all actors, he does qualify as an important hub.) In second place was Christopher Lee (2.684), followed by Dennis Hopper (2.698). Donald Sutherland ranked 4th. Among women, the most connected was Karen Black, in 21st place.

It's such highly connected actors, or hubs, that make it easy to get from any actor to any other in just a few steps. Choose any two actors at random. You can probably connect them in three steps or less. And it would be unusual to need more than four. If you search and search, you can find a few who would require more steps, but that is likely only if you deliberately choose actors who would not be very connected, like someone who appeared in only one film. (And remember, I said you need to choose two at random.)

So for example, just off the top of my head I'll say Basil Rathbone (because I saw a Sherlock Holmes movie last night) and Lindsay Lohan (no reason—I will not admit to having seen *Herbie: Fully Loaded*). These two actors are from entirely different eras, and Lohan is young and has been in relatively few films. But you can link them in only three steps. An aging Rathbone appeared in *Queen of Blood* (1966) with the pre–*Easy Rider* Dennis Hopper. Hopper was in *Knockaround Guys* with Bruce McFee, who appeared in *Confessions of a Teenage Drama Queen* (2004) with Lohan. The short path between Rathbone and Lohan was made possible by the hub provided by Hopper, who is, in fact, much more connected then Bacon.[7] (Hopper is connected directly to 3,503 other actors, about 1,500 more than Bacon.)

Hubs like Hopper make the actor network a small world. They make getting from one node to another easy, in much the way that major airport hubs like Chicago's O'Hare or Dallas–Fort Worth unite the smaller airports in airline networks so you can get from one town to another without too many plane changes.

Such large hubs generally do not exist, though, in either random or regular networks. In a regular network, every node has the same number of links, so there are no hubs. In a random network, shortcuts exist, but they might be very hard to find because prominent hubs would be very rare. In random networks, any one node (actor or airport) is as likely to be linked as much as any other, so most of them are linked to about the same degree. Only a few would have a lot more links than average, or a lot less. If actors were linked randomly, their rankings by number of links would form a bell curve, with most of them close to the middle. But in many small-world networks, there is no such "typical scale" of the number of links.

Such distributions—with no typical common size—are known as "scale free." In scale-free networks, many lonely nodes will have hardly any connections at all, some nodes will be moderately well connected, and a few will be superconnected hubs. To mathematicians and physicists, such a scale-free distribution is a sure sign of a "power law."

In a groundbreaking paper published in *Science* in 1999, Réka Albert and Albert-László Barabási of Notre Dame University noted the scale-free nature of many kinds of networks, and consequently the usefulness of power laws for describing them. The revelation that networks could be described by power laws struck a responsive chord among physicists. (They "salivate over power laws," Strogatz says—apparently because power law discoveries in other realms of physics have won some Nobel Prizes.)

Power laws describe systems that include a very few big things and lots of little things. Cities, for example. There are a handful of U.S. cities with populations in the millions, a larger number of medium-sized cities in the 100,000 to a million range, and many, many more small towns. Same with earthquakes. There are lots of little earthquakes, too weak to notice; a fewer number of middling ones that rattle the dishes; and a very few devastating shocks that crumble bridges and buildings.

In their *Science* paper, Barabási and Albert showed how the probability that a node in a scale-free network is linked to a given number of other nodes diminishes as the number of links increases. That is to say, scale-free networks possess many weakly linked nodes, fewer with a moderate number of links, and a handful of monsters—like Google, Yahoo, and Amazon on the World Wide Web. Nodes with few links are common, like small earthquakes;

nodes with huge numbers of links are relatively rare, like huge earthquakes. And the rate at which the probability of links goes down is quantified by a power law, just like the math describing the distribution of city or earthquake sizes. In other words, a good theory of networks should explain not only how Kevin Bacon (or Dennis Hopper) can be so connected, but also why networks are analogous to earthquakes.

Barabási and Albert proposed an explanation based on the recognition that networks are rarely static arrangements of nodes with fixed numbers of links, but rather are usually growing and evolving structures. As networks grow by adding new nodes, Barabási and Albert hypothesized, new links do not form at random. Rather each new node prefers to link to nodes that already have a lot of links. In other words, the rich get richer, and the result of such a growth process is a scale-free network with very rich hubs. The dynamics of the process indicated that "the development of large networks is governed by robust self-organizing phenomena that go beyond the particulars of the individual systems," Barabási and Albert noted.[8]

While their "preferential attachment" scheme did indeed predict the formation of hubs, it did not explain many other aspects of real-world networks, including clustering. And it turned out that not all small-world networks are scale-free. Barabási and Albert's original work, for instance, suggested that the networks explored by Watts and Strogatz were scale-free as well as being small worlds. But they aren't. The power grid is a small world but isn't scale-free, and neither is the neural network of *C. elegans*, Strogatz said. Still, there are many examples of networks that are both small-world and scale-free, with the World Wide Web being one spectacular example. And social networks *are* typically both small-world and scale-free, so understanding networks in terms of power laws would seem a good strategy for using networks to study human interactions.

Following Barabási and Albert's pioneering efforts to quantify network evolution, many other groups have joined the hunt to identify all the important qualities of networks and devise mathematical models to explain them. Among the organizations tuned into this issue is Microsoft, which obviously has a great interest in the Internet and World Wide Web. So their top scientists are busy investigating network math themselves. Leaders of this pack are a husband-wife team, mathematician Jennifer Tour Chayes and her husband/collaborator Christian Borgs. When I visited the Microsoft research labs outside Seattle, they outlined to me their efforts to identify the features that network math needs in order to capture the essence of the Web's structure.

"The Internet and the World Wide Web are grown, they're not engineered," Chayes pointed out. "No one really planned the Internet, and certainly no one planned the structure of the World Wide Web." Consequently the Web embodies many of the nuances of natural networks that a good mathematical model will need to capture, such as the small-world property (the ability to get from one page to any other in a relatively few number of steps) and the clustering phenomenon (if a Web page links to 10 others, there's a good chance that many of those 10 will link to one another as well). A further important feature is the preferential attachment identified by Barabási that conditions how a network grows, or ages. As the Web grows, and pages are added to the network, the older pages do tend to acquire more links than newcomers. But it's not always true that the oldest pages are the most connected. "It's not just a function of aging," Chayes explained. AltaVista, for example, once was the Cadillac of Web search engines. But the younger Google now has many more links. So different sites must earn links not only by virtue of age, but also beauty—or "fitness."

"AltaVista has been around longer but more people tend to link to Google—it's in some sense a better page," said Chayes. "All other things equal, the older sites will on average have more links, but if one site is more fit than another, that compensates for age. . . . If I'm twice as fit and I'm half as old, I should tend to have about the same number of connections."[9]

Another important feature of the Web, shared by many (but not all) networks, is that the links are "directed." Unlike the Internet, where wires run both ways, Web page hyperlinks go in only one direction. "Just because I link to CNN, that doesn't mean that CNN links to me," Chayes said. (Although, of course, it should.)

Naturally, good network math also needs to show that the Web is scale-free. A few sites have a huge number of links, more have a medium number, and most have very few links at all. Chayes and Borgs emphasized that equations describing the Web should predict not only such a distribution of links, but also the presence of a "strongly connected component" of Web pages. In the strongly connected component, or SCC, you can move from any page to another by following hyperlinks one page at a time. If her page is in the SCC, says Chayes, she can find a path to any other SCC page. "I can follow a series of hyperlinks and get to that person's page, and that person can follow a series of hyperlinks and get back to my page."

Borgs pointed out that some Web pages can link into the SCC even though no path links back to them. Some pages get linked *to* from a page in the SCC but don't link *back* to an SCC page. Knowing which pages are within the SCC, or connected to it in which way, would be important information for Web advertisers, he noted.

Building mathematical models that reproduce all these features of the Web is still a work in progress. But the models of the Web and other networks devised so far suggest that mathematicians of the future may someday be able to explain the behavior of many networks encountered in human affairs—such as economic, political and social networks; ecosystems; protein networks inside cells; and networks of contact that spread diseases. "I think there *is* going to be a mathematics of networks," said Chayes. "This is a very exciting new science."

BACK TO THE GAMES

Since game theory also claims dominion over describing human behavior, I asked Chayes whether it had any role to play in the new math of social networks. Fortunately, she said yes. "We are trying to explain why these network structures have evolved the way they have evolved, and that's really a game theory problem,"

Chayes said. "So there's a lot of work on game theory models for the growth of the Internet and the World Wide Web."

In fact, Chayes, Borgs, and collaborators have shown how math that is similar at least in spirit to a game theory approach can explain the emergence of preferential attachment in an evolving network (rather than merely assuming it, as Barabási and Albert did). It's a matter of minimizing the costs of competing considerations—the cost of making a connection, and the cost of operating it once it has been made. (It's kind of like buying a car—you can get a cheap one that will cost you more to keep running, or shell out more up front for high performance with low maintenance.) That trade-off can be viewed as a competition between different network structures, and the math that forecasts minimum cost also predicts that something like preferential attachment will describe the network's evolution.

More explicit uses of game theory have been called on to explain the evolution of other kinds of networks. A popular use of network models, for instance, is in making sense of the mess of chemicals interacting inside living cells. The interplay of thousands of proteins ends up determining how cells behave, which is often a matter of life and death. Game theory can help explain how those biochemical networks evolved into their current complex form.

Biologists would, of course, naturally assume that cellular metabolism should evolve to reach some "optimal" condition for fueling cell activity most efficiently. But what's most efficient? That depends on the environment, and the environment includes other species evolving toward optimality. "Thus, by evolving towards optimal properties, organisms change their environment, which in turn alters the optimum," note computational biologist Thomas Pfeiffer and biophysicist Stefan Schuster.[10] And that is just the sort of dynamic for which game theory—particularly evolutionary game theory—is optimal. For example, a key molecule in the network of cellular chemistry is ATP, which provides the energy needed to drive important metabolic processes. ATP is the product of a chain of chemical reactions. To stay alive, a cell needs a constant source of ATP, so the reaction "assembly" line has to operate 24/7.

There are, of course, different possible arrangements of the assembly line—that is, different combinations of reactions that could produce ATP (as with many networks, there being multiple pathways to get to a hub). An important question in cellular biology is whether cells should prefer to produce ATP as rapidly as possible, or as efficiently as possible (that is, with pathways that produce greater quantities of ATP from the same amount of raw material, getting more ATP bang for the buck). Some reaction pathways are faster but more wasteful than others, posing a trade-off for cells desiring to achieve an optimum metabolism.

The best strategy, a game-theoretic analysis shows, depends on the various other organisms in the vicinity competing for resources. Where competition is present, game theory recommends fast but wasteful ATP production, a prediction that contradicts straightforward notions of optimizing resource allocation. After all, if a population of microbial cells are competing for food, it would seem best for the group for each microbe to make the most efficient use of the available food supply, so there will be enough to go around. But game theory says otherwise—it's another example of the Prisoner's Dilemma in action. What's best for the individuals doesn't compute to be the overall best deal for the group.

"This paradoxically implies that the tendency of the users to maximize their fitness actually results in a decrease in their fitness—a result that cannot be obtained from traditional optimization," Pfeiffer and Schuster point out. "In the framework of evolutionary game theory, slow and efficient ATP production can be seen as altruistic cooperative behavior, whereas fast and inefficient ATP production can be seen as selfish behavior."[11]

But it's also a mistake to assume that cells will always act selfishly to enhance their survival odds. Game theory math suggests that in scenarios where a microbe's neighbors eat a different kind of food (so there is no competition for a single resource), more efficient production of ATP at the expense of speed would be a better survival strategy. Actual observations confirm that cells typically consuming resources shared by others (such as certain yeast cells) have evolved metabolisms that use the fast-but-wasteful approach to producing ATP. In multicellular organisms, though, cells behave more cooperatively with their neighbors, evolving reaction pathways that produce ATP more efficiently.

Intriguingly, cancer cells seem to violate the cooperation strategy and behave more selfishly (in terms of using inefficient ATP-producing processes). Game theory hasn't exactly cured cancer yet, but insights into such properties of cancer cells may contribute to progress in fighting it.

On a higher evolutionary level, a combination of network math and game theory may be able to explain more advanced forms of human cooperative behavior. Evolutionary game theory's assault on the cooperation problem—how altruistic behavior can evolve in societies of seemingly selfish individuals—has relied mainly on playing the Prisoner's Dilemma game under a variety of circumstances. In some versions of the game, the players (or agents) may encounter anybody else in the population and then decide whether to defect or cooperate. In one version, though, the agents face such decisions only in interactions with their immediate neighbors (the game, in other words, is "spatially structured"). It appears that cooperation is more likely to evolve in games with spatial constraints, at least when the game is the Prisoner's Dilemma.

But perhaps the Prisoner's Dilemma does not always capture the essence of real life very accurately. Life might sometimes more closely resemble a different kind of game. One candidate is the "snowdrift" game, in which the best strategic choice differs from the classic Prisoner's Dilemma. In a Prisoner's Dilemma, each player earns the highest payoff by defecting, regardless of what the other player does. In the snowdrift game, your best move is to defect only if your opponent cooperates. If the opponent defects, you are better off cooperating.[12] As it turns out, spatial constraints also influence the evolution of cooperation in the snowdrift game, but in a different way—inhibiting cooperation rather than enhancing it. That is a perplexing finding, calling into question game theory's validity for studying the cooperation issue.

However, as physicists Francisco Santos and Jorge Pacheco have pointed out, the "spatial constraint" of agents interacting only with their neighbors is not realistic, either. A more realistic spatial description of the agents, or players, is likely to be a scale-free network of the agent's relationships, simulating actual social connections. Merging the math of scale-free networks with game theory, the physicists found that cooperation ought to emerge with *either* the Prisoner's Dilemma or snowdrift games. "Contrary to previous results, cooperation becomes the dominating trait on both the Prisoner's Dilemma and the snowdrift game, for all values of the relevant parameters of both games, whenever the network of connections correspond to scale-free graphs generated via the mechanisms of *growth* and *preferential attachment*," the physicists reported in 2005 in *Physical Review Letters*.[13]

Numerous other papers have explored links between game theory and network math. It strikes me as a sensible trend that is bound to bear ever more mathematical fruit. Networks are, after all, complex systems that have grown and evolved over time. And game theory, as evolutionary biologists have discovered, is a powerful tool for describing the evolution of such complexity. (One paper specifically models a version of the Prisoner's Dilemma game showing how repeated play can lead to a complex network in a state that the authors refer to as a "network Nash equilibrium.")[14] Game theory's importance to society thus cannot help but expand dramatically as the critical nature of social networks becomes ever more clear.

In fact, physicists building their version of a Code of Nature with the tools of statistical mechanics (as did Asimov's Hari Seldon) have turned increasingly to using those tools on a network-based foundation. This alliance of statistical physics and network math, coupled with game theory's intimate links to networks, argues that game theory and statistical physics may together nourish the new science of collective human behavior that physicists have already begun to call sociophysics.

9

Asimov's Vision

Psychohistory, or sociophysics?

"Humans are not numbers." Wrong; we just do not *want* to be treated like numbers.

—Dietrich Stauffer

In 1951—the same year that John Nash published his famous paper on equilibrium in game theory—Isaac Asimov published the novel *Foundation*. It was the first in a series of three books (initially) telling the story of a decaying galactic empire and a new science of social behavior called psychohistory. Asimov's books eventually became the most famous science fiction trilogy to appear between *Lord of the Rings* and *Star Wars*. His psychohistory became the model for the modern search for a Code of Nature, a science enabling a quantitative description and accurate predictions of collective human behavior.[1]

Mixing psychology with math, psychohistory hijacked the methods of physics to forecast—and influence—the future course of social and political events. Today, dozens of physicists and mathematicians around the world are following Asimov's lead, seeking the equations that capture telling patterns in social behavior, trying to show that the madness of crowds has a method.

As a result, Asimov's vision is no longer wholly fiction. His psychohistory exists in a loose confederation of research enterprises that go by different names and treat different aspects of the issue. At various schools and institutes around the world, collaborators from diverse departments are creating new hybrid disciplines, with names like econophysics, socionomics, evolutionary economics, social cognitive neuroscience, and experimental economic anthropology. At the Santa Fe Institute, a new behavioral sciences program focuses on economic behavior and cultural evolution. The National Science Foundation has identified "human and social dynamics" as a special funding initiative.

Almost daily, research papers in this genre appear in scientific journals or on the Internet. Some examine voting patterns in diverse populations, how crowds behave when fleeing in panic, or why societies rise and fall. Others describe ways to forecast trends in the stock market or the likely effect of antiterrorist actions. Still others analyze how rumors, fads, or new technologies spread.

Diverse as they are, all these enterprises share a common goal of better understanding the present in order to foresee the future, and possibly help shape it. Put them all together, and Asimov's idea for a predictive science of human history no longer seems unthinkable. It may be inevitable.

Among the newest of these enterprises—and closest to the spirit of Asimov's psychohistory—is a field called sociophysics. The name has been around for decades, but only in the 21st century has it become more science than slogan. Like Asimov's psychohistory, sociophysics is rooted in statistical mechanics, the math used by physicists to describe systems too complex to expose the intimate interactions of their smallest pieces. Just as physicists use statistical mechanics to show how the temperature of two chemicals influences how they react, sociophysicists believe they can use statistical mechanics to take the temperature of society, thereby quantifying and predicting social behavior.

Taking society's temperature isn't quite as straightforward as it is with, say, gas molecules in a room. People usually don't behave the same way as molecules bouncing off the walls, except during some major sporting events. To use statistical physics to take society's temperature, physicists first have to figure out where to stick the thermometer.

Fortunately, the collisions of molecules have their counterpart in human interaction. While molecules collide, people connect, in various sorts of social networks. So while the basic idea behind sociophysics has been around for a while, it really didn't take off until the new understanding of networks started grabbing headlines.

Social networks have now provided physicists with the perfect playground for trying out their statistical math. Much of this work has paid little heed to game theory, but papers have begun to appear exploring the way that variants on Nash's math become important in social network contexts. After all, von Neumann and Morgenstern themselves pointed out that statistical physics provided a model giving hope that game theory could describe large social groups. Nash saw his concept of game theory equilibrium in the same terms as equilibrium in chemical reactions, which is also described by statistical mechanics. And game theory provides the proper mathematical framework for describing how competitive interactions produce complex networks to begin with. So if the offspring of the marriage between statistical physics and networks is something like Asimov's psychohistory, game theory could be the midwife.

SOCIOCONDEMNATION

Network math offers many obvious social uses. It's just what the doctor ordered for tracking the spread of an infectious disease, for instance, or plotting vaccination strategies. And because ideas can spread like epidemics, similar math may govern the spread of opinions and social trends, or even voting behavior.

This is not an entirely new idea, even within physics. Early attempts to apply statistical physics to such problems met with severe resistance, though, as Serge Galam has testified. Galam was a student at Tel-Aviv University during the 1970s, when statistical mechanics was the hottest topic in physics, thanks largely to some Nobel Prize–winning work by Kenneth Wilson at Cornell University. Galam pursued his education in statistical physics but with a concern—its methods were so powerful that all the important problems of inert matter might soon be solved! So he began to advocate the use of statistical physics outside physics, especially for analyzing human phenomena, and published several papers along those lines. He even published one with "sociophysics" in the title in 1982. The response from other physicists was not enthusiastic.

"Such an approach was strongly rejected by almost everyone," he wrote, "leading and non-leading physicists, young and old. To suggest humans could behave like atoms was looked upon as a blasphemy to both hard science and human complexity, a total non-sense, something to be condemned."[2]

My impression is that most physicists nowadays are not so hostile to such efforts (although some are) but are just mostly indifferent. There are some enthusiasts, though, and international conferences have been devoted to sociophysics and related topics. And thanks to the rapid advances in network math, the study of social networks has gained a certain respectability, diminishing the danger of instant condemnation for anyone pursuing it (although acceptance is clearly greater in Europe than in the United States).

Part of this acceptance probably stems from the growing popularity of an analogous discipline known as econophysics, a much more developed field of study. Econophysics[3] studies the interacting agents in an economy using statistical physics, and some prominent physicists have been attracted to it. Many young physicists have taken their skills in this field to Wall Street, where they can make money without the constant fear of government budget cuts.

Sociophysics is much more ambitious. It should ultimately encompass econophysics within it, along with everything else in the realm of human interactions. Of course, it has a way to go. But whatever anybody thinks of this research, there is certainly now a lot of it. Galam himself remains a constant contributor to the field. Now working in France, he has studied such social topics as the spread of terrorism, for instance, trying to identify what drives the growth of terrorism networks. In other work, he has analyzed opinion transmission and voting behaviors, concluding that "hung election scenarios," like the 2000 U.S. presidential contest, "are predicted to become both inevitable and a common occurrence."[4] Other researchers have produced opinion-spreading papers that try to explain whether an extreme minority view can eventually split a society into two polarized opposite camps, or even eventually become an overwhelming majority.

Most of this work is based on simple mathematical models that try to represent people and their opinions in a way that can be easily dealt with mathematically. There is no point in trying to be completely realistic—no amount of math could capture all the nuances in the process by which even a single individual formed his or her opinions, let alone an entire population. The idea is to find a simple way to represent opinions at their most basic and to identify a few factors that influence how opinions change—in a way that lends itself to mathematical manipulation. If the math then reproduces something recognizable about human behavior, it can be further refined in an attempt to inch closer to reality.

It's not hard to find people who think the whole enterprise is preposterous. Human beings are not particles—they bear not the slightest similarity to atoms or molecules. Why should you expect to learn anything about people from the math that describes molecular interactions?

On the other hand, molecules are not billiard balls—yet Maxwell made spectacular progress for physics by analyzing them as though they were. In his paper introducing statistical considerations to the study of gases, Maxwell applied his math to a system containing "small, hard and perfectly elastic spheres acting on one another only during impact." Molecules are small, to be sure, but otherwise that description is not very complete or accurate, as Maxwell knew full well. But he believed that insights into the behavior of real molecules might emerge by analyzing a simplified system.

"If the properties of such a system of bodies are found to correspond to those of gases," Maxwell wrote, "an important physical analogy will be established, which may lead to more accurate knowledge of the properties of matter."[5] Today, physicists hope to find a similar analogy between particles and people that will lead to an improved knowledge of the functioning of society.

SOCIOMAGNETISM

One popular example of such an approach appeared in 2000 from Katarzyna Sznajd-Weron of the University of Wroclaw in Poland. She was interested in how opinions form and change among members of a society. She reasoned that the global distribution of opinions in a society must reflect the behavior and interactions of individuals—in physics terms, the macrostate of the system must reflect its microstate (like the overall temperature or pressure of a container of gas reflects the speed and collisions of individual molecules).[6] "The question is if the laws on the microscopic scale of a social system can explain phenomena on the macroscopic scale, phenomena that sociologists deal with," she wrote.[7]

Sznajd-Weron was well aware that people recoil when told they are just like atoms or electrons rather than individuals with feelings and free will. "Indeed, we are individuals," she wrote, "but in many situations we behave like particles." And one of those common properties that people share with particles is a tendency to be influenced by their neighbors. Sometimes what one person does or thinks depends on what someone else is doing, just as one particle's behavior can be affected by other particles in its vicinity.

Sznajd-Weron related an anecdote about a New Yorker staring upward at the sky one morning while other New Yorkers pass by, paying no attention. Then, the next morning, four people stare skyward, and soon others stop as well, all looking up for no reason other than to join in the behavior of the crowd. Such pack behavior suggested to Sznajd-Weron an analogy for crowd behavior as described by the statistical mechanics of phase transitions, the sudden changes in condition such as the freezing of water into ice. Another sort of phase transition, of the type that attracted her attention, is the sudden appearance of magnetism in some materials cooled below a certain temperature.

It makes sense to relate society to magnetism, since society reflects the collective behavior of people, and magnetism reflects the collective behavior of atoms. A material like iron can be magnetic because its atoms possess magnetic properties, thanks to the arrangement of their electrons, the electrically charged fuzzballs that shield each atom's nucleus. Magnetism is related to the direction in which electrons spin. (You can view the spins as around an axis either pointing up or pointing down, corresponding to whether the electron spin is clockwise or counterclockwise.)

Ordinarily a bar of iron is not magnetic, because its atoms are directing their magnetism in random directions, so they cancel out. If enough atoms align themselves in one particular direction, though, others will follow—kind of like the way if enough people look up to the sky, everybody else will, too. When all the atoms line up, the iron bar becomes a magnet. It's as though each atom checks to see how its neighbor's electrons are spinning. When two atoms are sitting next to each other, their partnerless electrons want to spin in the same direction—that confers a slightly lower energy on the system, and all physical systems seek the state of lowest possible energy. Consequently the spin of one iron electron can influence the spin of its neighbor, inducing it to take on the same orientation. (In most materials an atom's electrons are mostly paired off with opposite spins. But iron and a few other materials possess some properly positioned electrons without partners. Magnetism is a little more complicated than this crude picture, of course, but the basic idea is good enough.)

As scientists began to understand this aspect of magnetism, they wondered if such local interactions between neighbors could explain the global phase transition from the nonmagnetic to magnetic state. In the 1920s, the German physicist Ernst Ising tried to show how neighboring spins could induce a spontaneous phase transition in a system, but failed. The problem was not in the basic idea, though—it was that Ising analyzed only a one-dimensional system, like a string of spinning beads on a necklace. Soon other researchers showed that Ising's approach did turn out to work when applied to two-dimensional systems, like spinning balls arranged in a grid.

Magnetism could thus be understood as a collective phenomenon stemming from the interactions of individuals—sort of like pack journalism. When one newspaper makes a big deal about a major story, all the other media jump in and beat the story to death—all the news is taken over by something like O.J. or Michael Jackson or some Runaway Bride. Similarly, rapid large-scale changes mimicking phase transitions occur in biology or the economy, such as mass extinctions or stock market crashes. In recent years it has occurred to physicists like Galam, Sznajd-Weron, and many others that the same principle could apply to social phenomena, such as the rapid spread of popular fads.

Sznajd-Weron set out to devise an Ising-like model of social opinions, trying out a very simplified approach that would be easy to handle mathematically. Instead of up or down spins, people could take a yes or no stance with respect to some issue. If you start out with opinions at random, how would the system change over time? Sznajd-Weron proposed a model based on the idea of "social validation." Just as the behavior of the New York skywatcher spread when others did the same, identical opinions between neighbors could cause their same opinion to spread socially, in a way similar to the way magnetism develops through Ising interactions.

Sznajd-Weron's model of society was pretty simple—something like one long street with houses on only one side. Each house is identified by a number (OK, that's realistic), and each house gets one opinion (or spin): either yes (mathematically represented as $+1$), or no (-1).

To start out, the houses all have opinions at random. Then, every day each house checks its neighbors and modifies its opinions based on some simple mathematical rules. Based on neighboring opinions, each house may (or may not) modify its own. In Sznajd-Weron's model, you start by considering two neighbors—let's say House 10 and House 11. Each of that pair has another neighbor (House 9 and House 12). Sznajd-Weron's rules say that if houses 11 and 12 have the same opinion, then houses 9 and 12 will adjust their opinions to match the common opinion of 10 and 11. If houses 10 and 11 disagree, though, House 9 will adjust its opinion to agree with House 11, and House 12 will change to agree with House 10.

Mathematically, the rules look like this, with S representing a house and the subscript i representing the house number (in the above example, S_i is House 10, S_{i+1} is House 11, etc.):

$$\text{If } S_i = S_{i+1} \text{ then } S_{i-1} = S_i \text{ and } S_{i+2} = S_i$$
$$\text{If } S_i = -S_{i+1} \text{ then } S_{i-1} = S_{i+1} \text{ and } S_{i+2} = S_i$$

In other words, when the two neighbors (10 and 11) agree, the two outside neighbors will share that opinion. If the first two neighbors disagree, then the one on the left will agree with the second and the one on the right will agree with the first. Why should that be? No reason, it's just a model. In a variant on Sznajd-Weron's original proposal, the second rule is switched:

$$\text{If } S_i = -S_{i+1} \text{ then } S_{i-1} = S_i \text{ and } S_{i+2} = S_{i+1}$$

In the original model, Sznajd-Weron performed computer simulations on a street with 1,000 houses and watched as opinions changed over 10,000 days or so. No matter how the opinions started out, the neighborhood eventually reached one of three stable situations—either all the houses voting yes, all no, or a 50-50 split. (These conditions correspond, in Sznajd-Weron's words, to either "dictatorship" or "stalemate.")

Since not all societies are dictatorships or stalemates, the model does not reflect the true complexity of the real world. But that doesn't mean the model is dumb—it means that the model has told us something, namely that more than just local interaction between neighbors is involved in opinion formation. And you don't need to know what all those other factors are to improve the model—you just need to know that they exist. In her 2000 paper,

Sznajd-Weron showed that such unknown factors (in technical terms, noise) could be described as a "social temperature" raising the probability that an individual would ignore the neighbor rules and choose an opinion apparently at random. With a sufficiently high social temperature, the system can stay in some disordered state, more like a democracy, rather than becoming a stalemate or dictatorship.

Even so, as Sznajd-Weron pointed out, her one-dimensional model is not likely to be very useful for social systems, just as Ising's one-dimensional model did not get the magnetism picture right, either. So in the years since her proposal, she and others have worked on extensions of the model. A similar model in two dimensions (with the "houses" occupying points on a grid) was developed by Dietrich Stauffer of the University of Cologne, probably today's most prominent sociophysicist. With the people aligned on a grid, everybody has four neighbors, a pair has six neighbors, and a block of four has eight neighbors. In this case, one rule might be that a block of four changes its eight neighbors only if all four in the block have the same spin (or opinion). Or two neighbors paired with the same spin can change the spins of their six neighbors. A grid model can accommodate more complications and thus reproduce more of the real properties of society.

SOCIONETWORKS

Clearly, though, the way to get more social realism is to apply such rules not to simple strings or grids but to the complex social networks discovered in the real world. And much interesting work has begun to appear along these lines. One approach examines the general idea of "contagion"—the spread of anything through a population, whether infectious disease or ideas, fads, technological innovations, or social unrest. As it turns out, fads need not always spread the same way as a disease, as different scenarios may guide the course of different contagions.

In some cases, a small starting "seed" (a literal virus, perhaps, or just a new idea) can eventually grow into an epidemic. In other cases a seed infects too few people and the disease or idea dies out. Peter Dodds and Duncan Watts (of the Watts-Strogatz network paper) of Columbia University have shown that what happens can depend on how much more likely a second exposure is to infect an individual than a first exposure. Their analysis suggests that the spread of diseases or ideas depends less on "superspreaders" or opinion leaders than on how susceptible people are—how resistant they are to disease or how adamantly they hold their current opinion. Such results imply that the best way to hamper or advance contagion would be strategies that increase or reduce the odds of infection. Better health procedures, for instance, or financial incentives to change voting preferences, could tip the future one way or another.

"Our results suggest that relatively minor manipulations . . . can have a dramatic impact on the ability of a small initial seed to trigger a global contagion event," Dodds and Watts declared in their paper.[8] It sounds like just the sort of thing that Hari Seldon incorporated into psychohistory, so that his followers could subtly alter the course of future political events.

In real life, of course, people don't necessarily transmit opinions or viruses in the simple ways that such analyses assume. So some experts question how useful the statistical mechanics approach to society will ultimately be. "I think in some limited domains it might be pretty powerful," says Cornell's Steven Strogatz. "It really is the right language for discussing enormous systems of whatever it is, whether it's people or neurons or spins in a magnet. . . . But I worry that a lot of these physicist-style models of social dynamics are based on a real dopey view of human psychology."[9]

Of course, that is precisely where game theory comes into play. Game theory has given economists and other social scientists the tool for quantifying human psychology in ways that Freud could only dream of. Neuroeconomics and behavioral game theory have already sculpted a much more realistic model of human psychology than the naive *Homo economicus* that lived only to maximize money. And once you have a better picture of human psychology—in particular, a picture that depicts the psychological variations among individuals—you need game theory to tell you what happens when those individuals interact.

SOCIOPHYSICS AND GAME THEORY

After all, when you get to really complex social behaviors—not just yes or no votes, but the whole spectrum of human cultural behavior and all its variations—the complex interactions between individuals really do matter. It is yet again similar to the situation with molecules in a gas. In his original math describing gas molecules, Maxwell considered their only interaction to be bouncing off of each other (or the container's walls), altering their direction and velocity. But atoms and molecules can interact in more complicated ways. Electrical forces can exert an attractive or repulsive force between molecules, and including those forces in the calculations can make statistical mechanical predictions more accurate.

Similarly, the behavior of people depends on how they are affected by what other people are doing, and that's what game theory is supposed to be able to describe. "Game theory was created," Colin Camerer points out, "to provide a mathematical language for describing social interaction."[10] Numerous efforts have been made to apply game theory in just that way. One particularly popular game for analyzing social interaction is the minority game, based on an economist's observations about a Santa Fe bar.

Keep in mind that in game theory, a player's choices should depend on what the other players are choosing. So the game as a whole reflects collective behavior, possibly described by a Nash equilibrium. In simple sociophysics models based on neighbors interacting, the global collective behavior results from purely local influences. But the Nash equilibrium idea suggests that individual behavior should be influenced by the totality of all the other behaviors. It may be, for instance, that the average choices of all the other players is the most important influence on any one individual's choice (in physics terms, that would correspond to a "mean-field theory" version of statistical mechanics).

In traditional game theory, each player supposedly is 100 percent rational with total information and unlimited mental power to figure out what everybody else will do and then calculate the best move. But sometimes (actually, almost all the time) those conditions are not satisfied. People have limited calculating power and limited information. There are situations where the game is too complex and too many people are involved to choose a foolproof decision using game theory.

And in fact, many simple situations can prove too complicated to calculate completely, even something as innocent as deciding whether to go to a bar on Friday night or stay home instead. This problem was made famous by Brian Arthur, an economist at the Santa Fe Institute, in the early 1990s. A Santa Fe bar called El Farol had become so popular that it was no longer always a pleasant place to go because of the crowds. (It was reminiscent of baseball player Yogi Berra's famous comment about the New York City restaurant Toots Shor's. "Toots Shor's is so crowded," said Yogi, "nobody goes there anymore.")

Arthur saw in the El Farol situation a problem of decision making with limited information. You don't know in advance how many people will go to the bar, but you assume that everybody would like to go unless too many other people are going also. Above some level of attendance, it's no fun. This situation can be framed as a game where the winners are those in the minority—you choose to go or stay home and hope that the majority of people make the opposite choice.

In 1997, Damien Challet and Yi-Cheng Zhang developed the mathematics of the El Farol problem in detail, in the form of what they called the minority game. Since then it has been a favorite framework of many physicists for dealing with economic and social issues.[11]

In the basic version of the game, each would-be bar patron (in mathematical models, such customers are called "agents") possesses a memory of how his or her last few bar-going decisions have turned out. (Players find out after every trial whether the stay-at-home or bar-going choices were the winners.) Suppose that Friday is your regular drinking night, and you can remember what happened three weeks back. Say, for instance, that on each of the last three Friday nights, a majority of people went to the bar. They were therefore the losers, as the minority of players avoided the crowd by staying at home. Your strategy for next Friday might be to go to the bar, figuring that after three loser trips in a row most people will decide to stay home and the bar will be less crowded. On the other hand, your strategy might be to go based only on the results of the past week, regardless of what happened the two weeks before.

At the start of the game, each agent gets a set of possible strategies like these, and then keeps track of which strategies work better than others. Over time, the agents will learn to use the strategies that work the best most often. As a result, the behavior of all the players becomes coordinated, and eventually attendance at the bar will fluctuate around the 50-50 point—on some Fridays a minority will go to the bar, and on some a slight majority, but attendance will never be too far off from the 50-50 split.

You don't have to be a drinker to appreciate the usefulness of the minority game for describing social situations. It's not just about going to bars—the same principles apply to all sorts of situations where people would prefer to be in a minority. You can imagine many such scenarios in economics, for instance, such as when it's better to be a buyer or a seller. If there are more sellers than buyers, you have the advantage if you're a buyer—in the minority.

Further work on the minority game has shown that in some circumstances it is possible to predict which choice is likely to be in the minority on next Friday night. It depends on how many players there are and how good their memory is. As the number of players goes down (or their memory capacity goes up), at some point the outcome is no longer random and can be predicted with some degree of statistical confidence.

MIXED CULTURES

While the minority game provides a good example of using (modified) game theory to model group behaviors, it still leaves a lot to be desired. And it certainly is a far cry from Asimov's psychohistory. Psychohistory quantified not only the interactions between individuals in groups, but also the interactions among groups, exhibiting bewildering cultural diversity. Today's nonfictional anthropologists have used game theory to demonstrate such cultural diversity, but it's something else again to ask game theory to explain it. Yet if sociophysics is to become psychohistory, it must be able to cope with the global potpourri of human cultural behaviors, and achieving that goal will no doubt require game theory.

At first glance, the prospects for game theory encompassing the totality of cultural diversity seem rather bleak. Especially in its most basic form, the ingredients for a science of human sociality seem to be missing. People are not totally rational beings acting purely out of self-interest as traditional game theory presumes, for example. Individuals playing games against other individuals make choices colored by emotion. And societies develop radically different cultural patterns of collective behavior. No Code of Nature dictates a universal psychology that guides civilizations along similar cultural paths.

As Jenna Bednar and Scott Page of the University of Michigan have described it, game theory would seem hopeless as a way to account for the defining hallmarks of cultural behavior. "Game theory," they write, "assumes isolated, context-free strategic environments and optimal behavior within them."[12] But human cultures aren't like that. Within a culture, people behave in similar, fairly consistent ways. But behavior differs dramatically from one culture to the next. And whatever the culture, behavior is typically not optimal, in the sense of maximizing self-interest. When incentives change, behavior often remains stubbornly stuck to cultural norms. All these features of culture run counter to some basic notions of game theory.

"Cultural differences—the rich fabric of religions, languages, art, law, morals, customs, and beliefs that diversifies societies—and their impact would seem to be at odds with the traditional game theoretic assumption of optimizing behavior," say Bednar and Page. "Thus, game theory would seem to be at a loss to explain the patterned, contextual, and sometimes suboptimal behavior we think of as culture."[13]

But game theory has a remarkable resilience against charges of irrelevance. It's explanatory power has not yet been exhausted, even by the demands of explaining the many versions of human culture. "Surprisingly," Bednar and Page declare, "game theory is up to the task."[14]

The individuals, or agents, within a society may very well possess rational impulses driving them to seek optimum behaviors, Bednar and Page note. But the effort to figure out optimal behaviors in a complicated situation is often considerable. In any given game, a player has to consider not only the payoff of the "best" strategy, but also the cost of calculating the best moves to achieve that payoff. With limited brain power (and everybody's is), you can't always afford the cost of calculating the most profitable response.

Even more important, in real life you are never playing only one game. You are in fact engaging in an ensemble of many different games simultaneously, imposing an even greater drain on your brain power. "As a result," write Bednar and Page, "an agent's strategy in one game will be dependent upon the full ensemble of games it faces."

So Alice and Bob (remember them?) may be participating in a whole bunch of other games, requiring more complicated calculations than they needed back in Chapter 2. If they have only one game in common, the overall demand on their calculating powers could be very different. Even if they face identical situations in the one game they play together, their choices might differ, depending on the difficulty of all the other games they are playing at the same time. As Bednar and Page point out, "two agents facing different ensembles of games may choose distinct strategies on games that are common to both ensembles."

In other words, with limited brain power, and many games to play, the "rational" thing to do is not to calculate pure, ideal game theory predictions for your choices, but to adopt a system of general guidelines for behavior, like the Pirate's Code in the Johnny

Depp movie *Pirates of the Caribbean*. And that's what it means to
behave culturally. Cultural patterns of behavior emerge as individuals tailor a toolkit of strategies to apply in various situations,
without the need to calculate payoffs in detail. "Diverse cultures
emerge not in spite of optimizing motivation," Bednar and Page
write, "but because of how those motivations are affected by
incentives, cognitive constraints, and institutional precedents.
Thus agents in different environments may play the same game
differently."[15]

The Michigan scientists tested this idea with computer simulations on a variety of games, giving the agents/players enough
brain power to compute optimal strategies for any given game. In
the various games, incentives for the self-interested agents differed,
to simulate different environmental conditions. These multiple-
game simulations show that game theory itself drives self-
interested rational agents to adopt "cultural" patterns of behavior.
This approach doesn't explain everything about culture, of course,
but it shows how playing games can illuminate aspects of society
that at first glance seem utterly beyond game theory's scope.
And it suggests that the scope of sociophysics can be grandly
expanded by incorporating game theory into its statistical physics
formulations.

In any event, recent developments in the use of statistical physics in describing networks and society—and game theory's intimate relationship with both—instill a suspicion that game theory
and physics are somehow related in more than a superficial way. As
game theory has become a unifying language for the social sciences, attempts by physicists to shed light on social science inevitably must encounter game theory. In fact, that's exactly what has
already happened in economics. Just yesterday, the latest issue of
Physics Today arrived in my mail, with an article suggesting that
economics may be "the next physical science."

"The substantial contribution of physics to economics is still in
an early stage, and we think it fanciful to predict what will ultimately be accomplished," wrote the authors, Doyne Farner and
Eric Smith of the Santa Fe Institute and Yale economist Martin

Shubik. "Almost certainly, 'physical' aspects of theories of social order will not simply recapitulate existing theories in physics."[16]

Yet there are areas of overlap, they note, and "striking empirical regularities suggest that at least some social order . . . is perhaps predictable from first principles." The role of markets in setting prices, allocating resources, and creating social institutions involves "concepts of efficiency or optimality in satisfying human desires." In economics, the tool for gauging efficiency and optimality in satisfying human desires is game theory. In physics, analogous concepts correspond to physical systems treated with statistical mechanical math. The question now is whether that analogy is powerful enough to produce something like Asimov's psychohistory, a statistical physics approach to forecasting human social interaction, a true Code of Nature.

One possible weakness in the analogy between physics and game theory, though, is that physics is more than just statistical mechanics. Physics is supposed to be the science of physical reality, and physical reality is described by the weird (yet wonderful) mathematics of quantum mechanics. If the physics–game theory connection runs deep, there should be a quantum connection as well. And there is.

10

Meyer's Penny

Quantum fun and games

Do games have anything deeper to say about physics, or vice versa? Maybe. Most surprisingly, the connection might arise at the most fundamental level of all: quantum physics.

—Chiu Fan Lee and Neil F. Johnson, *Physics World*

It's the 24th century, aboard the starship *Enterprise*.

Captain Jean-Luc Picard places a penny heads up in a box, so that it can be touched but not seen. His nemesis Q, an alien with mysterious powers, then chooses whether to flip the coin over or not. Without knowing what Q has done, Picard then must decide to flip, or not flip, the coin as well. Q then gets the last turn. He either flips the penny or leaves it alone. If the penny shows heads when the box is removed, Q wins; tails wins for Picard.

They play the game 10 times, and Q wins them all.

It's not a scene from any actual episode of *Star Trek: The Next Generation*, but rather a scenario from a physics journal introducing an entirely new way of thinking about game theory.

The penny-flipping game is an old game theory favorite. It appears in various disguises, such as the game of chicken. (Whether you flip the coin or leave it alone corresponds to veering out of the path of the oncoming car or continuing straight on.) If they were playing the original version of the penny game, Q and Picard should, in the long run, break even, no one player winning more often than the other. Ten wins in a row for one player defies any reasonable definition of luck.

So if this had really happened on the show, Commander Riker would have immediately accused Q of cheating. But the wiser Picard would have pondered the situation a little longer and eventually would have realized that Q's name must be short for quantum. Only someone possessing quantum powers can always win the penny game.

As it turns out, Earth's physicists did not need an alien to teach them about quantum games. They emerged three centuries early, on the eve of the 21st century, out of an interest in using the powers of quantum mechanics to perform difficult computations. It was an unexpected twist in the story of game theory, as quantum games disrupted the understanding of traditional "classical" games in much the way that quantum mechanics disturbed the complacency of classical physics. The invention of quantum game theory suggested that the bizarre world of quantum physics, once restricted to explaining atoms and molecules, might someday invade economics, biology, and psychology. And it may even be (though perhaps not until the 24th century) that quantum games will cement the merger of game theory and physics. In fact, if physics ever finds the recipe for forecasting and influencing the social future, it might be that quantum game theory will provide the essential ingredient.

Now, if you've been reading carefully all this time, it might seem a little unfair that, after coming to grips with the complexities of game theory, network math, and statistical mechanics, you must now face the bewildering weirdness of quantum physics on top of it. Fortunately, the space available here does not permit the presentation of a course in quantum mechanics. Besides, you don't need to know everything there is to know about quantum physics to see how quantum game theory works. But you do have to be willing to suspend your disbelief about some of quantum theory's strangest features—most importantly, the concept of multiple realities.

QUANTUM TV

I've described this quantum confusion before (in my book *The Bit and the Pendulum*) by relating it to television. In the old days, TV signals traveled through the air, all the possible channels passing through your living room at the same time. (Nowadays they usually arrive via cable.) By turning the dial on your TV set (or punching a button on the remote control), you can make one of those shows—one of many possible realities—come to life on your screen. The realm of atoms, molecules, and particles even smaller works in a similar way. Left to themselves, particles buzz about like waves, and their properties are not sharply defined. In particular, you cannot say that a particle occupies any specific location. An atom can literally be in two places at once—until you look at it. An observation will find it located in one of the many possible positions that the quantum equations allow.

An important issue here, one that has occupied physicists for decades, involves defining just what constitutes an "observation." In recent years, it has become generally agreed that humans are not necessary to perform an observation or measurement on a particle. Other particles bouncing off it can accomplish the same effect. That is to say, an atom, on its own, cannot be said to occupy a specific location. But once other atoms start hitting it, the atom will become localized in a position consistent with the altered paths of the other atoms. This phenomenon is known as decoherence. As long as decoherence can be avoided (for example, by isolating a particle from other influences, maintained at very low temperatures), the weird multiplicity of quantum realities can be sustained.

This feature of quantum physics has been an endless source of controversy and consternation for physicists and nonphysicists alike. But experimental tests have left no room for doubt on this point. In the subatomic world, reality is fuzzy, encompassing a multiplicity of possibilities. And those possibilities all have a claim to being real. It's not just that you don't know where an atom is—it occupies no definite location, but rather occupies many locations simultaneously.

From a game theory point of view, there is a simple enough way of looking at this—reality itself is a mixed strategy.

Personally, I find this to be an uncanny analogy. In game theory, your best strategy typically is not one predetermined move or set of moves, but a mix of several strategies chosen with some particular probability—say, Strategy A 70 percent of the time and Strategy B 30 percent of the time. In the math of quantum physics, the location of a particle cannot be definitively determined, but only described probabilistically—perhaps you will find it in Region A 70 percent of the time, and in Region B 30 percent of the time. Still, at first glance, you wouldn't expect this analogy to be very meaningful. There's no reason to believe that the math of molecules would be relevant to making choices in economic games. But it turns out that applying quantum math to game theory does allow new decision-making strategies, adding a whole extra dimension to game theory's powers.

To be sure, some experts have doubted that quantum games offer any real benefits that could not be obtained in other ways. But other researchers have suggested that understanding quantum games could have ramifications for managing auctions, choosing better stock portfolios, and even improving the principles underlying democratic voting. And new technologies have begun to make experimental demonstrations of quantum games possible.

VON NEUMANN RETURNS

When you think about it, marrying quantum math to game theory is natural enough. In a way, the surprise is that nobody did it sooner. After all, John von Neumann, the originator of modern game theory, was also a pioneer in quantum mechanics. And the initial impetus for quantum games stemmed from the fact that von Neumann was also a pioneer in developing digital computers.

When David Meyer, a physicist-turned-mathematician at the University of California, San Diego, was invited to give a talk about quantum computing at Microsoft in January 1998, his thoughts turned to von Neumann. "I was giving a talk to the whole research division and I wanted to come up with something new to talk about," Meyer told me when I visited him at his office on the UCSD campus in La Jolla. "So I was thinking, what could I talk about at Microsoft that they'll be interested in?"[1]

Meyer's research had been focusing on quantum versions of computing, and he was naturally familiar with the fact that the standard version of quantum physics math had been developed by von Neumann. "And of course von Neumann is also responsible for the architecture of modern computers to a large extent, so that's relevant to Microsoft also," Meyer noted. "But then there's a third thing that von Neumann is known for—his invention of game theory, which is a big part of economics, and of course that's relevant to Microsoft also. So I thought, OK, how can I put these things together?" It seemed obvious that the thing to do was explore the possibility of a quantum version of game theory.

Meyer found a place to begin simply by considering game theory terminology. Von Neumann had shown that in two-player zero-sum games each player could always have a "best" strategy, but that it was not always a pure strategy—making the same play (for given circumstances) every time. In some cases the best strategy is to choose from various pure strategies with a certain probability for each—that is, a probability distribution of strategies, or "mixed" strategy.

"Now the fact that it's called a mixed strategy versus a pure strategy is not an accident," Meyer pointed out. "Von Neumann's responsible, as far as I can tell, for that vocabulary, and that vocabulary is the same vocabulary as in quantum mechanics. You have pure states and you have mixed states—mixed states are probability distributions over pure states. It has the same meaning."

So Meyer's talk at Microsoft explored a way of bringing quantum theory's multiple "mixed state" realities to game theory. He wisely chose one of the simplest games possible, the penny-flipping game. It's a simple game where the idea is simply to outguess your opponent, since there is no particular logic involved in deciding whether to flip the penny or not. If, however, one player discerns a pattern in the choices of the other, that knowledge could be exploited in repeated sessions of the game.

In a nonquantum or "classical" version of the game, Picard's best strategy would be to flip the coin half the time (in other words, he should flip a coin to decide whether to flip the coin), thereby making sure there will be no pattern to detect. Q, who gets two moves, should choose each of his four possible strategies one-fourth of the time (flip on both moves, flip on neither move, flip on the first move only, or flip only on the second move). If both players observe those strategies, they should each win half the time. Neither could do better by changing strategies, so it represents a Nash equilibrium.

In Meyer's quantum scenario, Picard still must play classically. But Q is allowed to play a quantum strategy—that is, he can flip the coin not from heads to tails but into a quantum combination of both possibilities, a coin that is half-heads and half-tails, like an electron that is simultaneously here and there.

In the lingo of quantum information physics, such a dual-valued head-tail combination is known as a qubit—short for a "quantum bit" of information. In traditional computing, bits are units of information corresponding to one of two possibilities—yes or no, heads or tails, 1 or 0. A classical coin must fall either heads or tails, but a quantum coin is permitted multiple possibilities, a mix of heads and tails at the same time. (I like to think of a qubit as a tossed coin while still spinning—it is neither heads nor tails until it is observed, sort of the way you don't know what a coin will show until it is caught or hits the ground.)

In actual quantum information experiments, the "coin" is usually a particle of light—a photon—and heads or tails might correspond to how the photon is spinning (the direction in which its axis of spin is pointing). For practical reasons, such experiments more often rely on measuring the photon's polarization, the orientation of the light wave (or more technically, of the electric field component of the light wave). Filters (like the polarized lenses of sunglasses) can block or transmit polarized light depending on its orientation, usually designated as horizontal or vertical. If you imagine the filter as something like a picket fence, a vertically polarized photon would pass through; a horizontally polarized photon would be blocked. (Of course, you could also transmit a photon oriented in between vertical and horizontal—that is, tilted. In that case, the recipient of the photon could tilt the detector, too, and block a photon tilted to the left by tilting the detector to the right.)

For Meyer's penny, turning it to heads or tails corresponds to setting the orientation of a polarizing filter—showing the head, say, but hiding the tail side of the coin.

Meyer's math showed how quantum manipulation of the penny can always guarantee that it will end up heads—a win for Q. Since Q plays first, he can use his quantum magic to flip the coin into a 50-50 mix of heads and tails. (In this case, rather than thinking of it as a spinning coin, you could imagine the penny standing on edge.) Therefore it doesn't matter what Picard does on his next move. Whether he flips or not, the penny remains on its edge (mathematically speaking). Q can then perform a reverse quantum move that will return the coin to its original condition—heads up.

If you'd like a more rigorous explanation, the coin-flipping game can be described in terms of the direction of the quantum coin's spin in a three-dimensional coordinate system (with the coordinate axes labeled x, y, and z). If you define heads as a spin pointing north on the z axis (in the "$+z$" direction), then tails would be pointing the opposite way (south, or the $-z$ direction). A classical flip (the only move allowed to Picard) merely switches the direction of the spin, from $+z$ to $-z$. Q, however, can perform a quantum twist to the spin, pointing it "east" (along the $+x$ axis). If Picard then flips the coin from north to south, the spin still points east, so it doesn't matter what Picard does—Q's next move returns the spin to north, or heads, and Picard loses.[2] Picard's strategy of flipping the coin half the time—guaranteed by classical game theory to be the best strategy he can play—turns out to be a worthless strategy against a quantum player.

There's a significance to this point that is easy to miss. Game theory supposedly tells you the best strategy you can play in a game. Meyer's discovery applies a huge asterisk to that statement. The footnote reads, in bold type, that game theory tells you the best strategy *only if you are able to ignore the multiple possibilities of quantum physics*. And since the world is, in fact, governed by quantum physics, it's more than possible that, in some circumstances at least, quantum games may someday be relevant to real-life situations.

QUANTUM DILEMMAS

Meyer's paper reporting the substance of his Microsoft talk was published in *Physical Review Letters* in 1999.[3] Soon thereafter, a second version of a quantum game (focusing on the famous Prisoner's Dilemma) appeared independently of Meyer's work. And in the next few years, dozens of other papers began to explore a whole gamut of quantum games. Most suggested that the outcomes in standard games, such as the Prisoner's Dilemma, might be improved with quantum strategies. Some papers applied quantum game principles to economics, suggesting, for instance, that the multiple possibilities of quantum physics might be applied to selecting the best mix of stocks in a portfolio, or in making decisions about when and whether to buy or sell.

It seems, though, that the original arguments that quantum strategies permit better outcomes in many games were not airtight. In some cases, merely allowing a "referee" to mediate between players, without any quantum magic, can achieve the same effects. If that were so, there would be nothing really inherently "quantum" about the games—they would just be different games, still classical, but with new rules. After thinking this through, however, Meyer concluded that there are still ways to make games truly quantum in character. "It's true that there are some aspects of quantum games which you can simulate by adding classical communication to what the players do," he told me. "But once you start thinking about adding classical communication . . . then for a fair comparison the quantum game should really be thought of as adding quantum communication to what the players do, and then there can be a difference." In other words, if the mediators or players are allowed to use quantum communication systems, quantum benefits may indeed be achieved.

"It's not that hard now to send quantum bits from one place to another," Meyer said. "So it's not implausible that you could . . . have players in some sort of game-theoretic setting, and you have the referee, rather than classical information, sending them quantum information—the advantage being that the outcome is different and possibly better."[4]

If so, he said, various real-life problems might be addressed with quantum game theory. Ways to make online voting both anonymous but verifiable, for example, might be possible with quantum information. Combinatorial auctions, such as the bidding by many companies for various licenses to be issued by the government, could perhaps be managed more efficiently by using quantum information to coordinate the bids.

"It's quite conceivable to me that some of these things may be doable in a better way or at least a different way by exchanging quantum information," Meyer said. "There's a huge realm to play in here. It's something that should be explored . . . and it might even be practical at some point."

QUANTUM COMMUNICATION

Quantum communication is, in fact, already feasible on a small scale, using optical fibers to transmit particles of light carrying qubits, the quantum bits of information. Qubits can be used to transmit secret codes with uncrackable quantum protection from eavesdroppers, guaranteeing that the code cannot be intercepted without detection. Quantum signaling of this nature has been demonstrated through several miles of optical cable and even through open air. Quantum-coded signaling to military satellites is well within the realm of technological possibility within the future timeframe of Pentagon budget planning.

To be practical, though, the more grandiose quantum game

schemes would probably need a tool that is today only in the infancy of its development—a quantum computer. In fact, one of the neatest things about quantum games is that they could give quantum computers something to do.

At the moment, quantum computers don't exist in any meaningful sense, although laboratory demonstrations of rudimentary quantum computation have been accomplished. If they can be scaled up to a useful size, quantum computers could exploit the multiple quantum realities to do many calculations simultaneously, thereby drastically shortening the time it takes to solve some very hard problems. So in theory, quantum computers could be enormously more powerful than today's supercomputers, but only for those special sorts of problems that lend themselves to quantum treatment. Massive database searching could be faster with a quantum computer, for example, and breaking secret codes is certainly something you wouldn't want to try without one.

Today's secret codes, used in military, financial, and other sorts of confidential communications, rely on the difficulty of breaking large numbers down into their prime factors. For small numbers, that's easy to do: 15, for instance, is obviously the product of the two prime numbers 5 and 3; 35 is the product of the primes 5 and 7. But for a number, say, 200 digits long, the world's fastest supercomputer might churn for a billion years without discovering the two primes that were multiplied to produce it. The way secret coding systems are set up, you can encode a message if you know the long number, but decode it only if you know its two prime factors.

This system seemed pretty secure, as it's unlikely anybody will care if your code gets broken a billion years from now. But in 1994, the mathematician Peter Shor showed that the primes could be discovered quickly—if you had a quantum computer at your disposal. The quantum computer could be programmed to explore all the prime possibilities at once; all the wrong answers would cancel themselves out, leaving a number that could be used to compute the primes easily. Designing and building a quantum computer is much simpler on paper than in practice, though, and it will doubtless be decades before you'll be able to buy one at Best Buy.

Nevertheless, simple quantum computations can be performed now. In fact, factoring 15 has been achieved with a quantum computer based on the same technology used in MRI medical imaging. And in 2002 Chinese physicists reported an experimental demonstration of the quantum Prisoner's Dilemma game, using a simple quantum computer. In the following year, Chinese physicists Lan Zhou and Le-Man Kuang outlined how to set up a quantum game communication system using lasers and mirrors and other optical devices in a paper published in *Physics Letters A*.[5]

QUANTUM ENTANGLEMENT

The Zhou-Kuang design exploits one of the most mysterious features of quantum physics, a ghostly bond between particles that have emerged from a mutual interaction. When two particles of light (photons) are emitted simultaneously from the same atom, for example, they retain an ethereal connection—measuring one seems to affect the other even if it is meters, miles, or light-years away. This connection is called "entanglement," and it's one of the things about quantum mechanics that really bothered Einstein (he called it "spooky action at a distance").

When two photons are entangled, they share quantum information in a peculiar way. If you think of them as spinning coins, they both keep spinning—becoming neither heads nor tails—until one of them is observed. And then the other one stops spinning, too! So suppose I possess two pennies, spinning within opaque boxes, entangled in such a way that if one is observed to be heads, the other will turn up tails. I send one box via FedEx to my sister in Ohio. She can't resist opening the package right away and discovers the penny on the bottom of the box, showing heads. The instant she sees it, the penny in my remaining box stops spinning and shows tails—whether I'm in Texas or California or on the International Space Station. Even if I don't look in my box, I know damn well that the penny shows tails (once my sister has called to tell me that hers is heads).[6] Somehow, my sister's observation of her penny influenced the state of my penny, no matter how far away we are. The same thing happens for real with actual photons of light, when you measure not heads or tails but how the photon is spinning or the orientation of its polarization.

This shared information between entangled particles can be exploited for many kinds of quantum communication purposes. In a quantum game, entangled particles can carry information about a player's choice in such a way that one choice will influence another. Take the Prisoner's Dilemma game. In the classical game, the players typically choose to defect because they cannot be sure that their partner will cooperate. Overall, the pair's best strategy is for both to remain silent—that way they'll serve the least amount of time. But each individual prisoner's best strategy is to rat out the other (to avoid the risk of a much longer sentence). So the best choice for the individual turns out not to be the wisest choice for the team. "We have a dilemma," write quantum game theorists Adrian Flitney and Derek Abbott, "some form of which is responsible for much of the misery and conflict throughout the world."[7]

But suppose there was a way that both players could make their decision hinge on the decision of the other. That's what playing with entangled photons offers. As Zhou and Kuang showed, you can set up an apparatus that allows the choice of "defect" (rat out your partner) or cooperate (keep mum) to be transmitted by photons, passing through a maze of mirrors and other optical devices, ultimately to reach a detector signifying either defect or cooperate. You can shoot your photon into the maze in different ways—so that it would end up in the "defect" detector, for instance, or in the "cooperate" detector. There's nothing tricky about that. But the maze can be set up so that the photons from the two players become entangled, with the result that *both* will end up cooperating. That is, you can send your choice in such a way that your photon will send a signal to cooperate only if the other one does, too.

This work shows that quantum game theory, at least in principle, could be used to alter in a deep and profound way the choices people make given the choices of others. Imagine, for instance, a quantum version of the "public goods" game discussed in earlier chapters. The idea is that a neighborhood group proposes building a project of public benefit, such as a park, to be paid for by voluntary contributions. Presumably people who want the park will contribute the most money to the fund drive. But standard game theory suggests that many people who want the park will contribute little or no money, reasoning that others will fork over enough to pay for it. Therefore it's hard to get donations, even for a park everybody desires, without the intervention of some outside agency (say, a tax collector).

In 2003, scientists from HP Labs in Palo Alto, California, posted a paper on the Internet showing how a quantum public goods game provides strategies that reduce the temptation to freeload. When people make economic or social decisions, they don't always choose based on self-interest alone, but may be influenced by social norms and expectations—sort of the way properties of a photon are influenced by distant measurements. So if you send your pledge via a quantum information channel, its message can depend on the messages from the other contributors. Therefore, the HP scientists suggested, entangled photons transmitted by laser beams through optical fibers could in theory be used for pledging donations in real-life community projects. Using quantum-entangled photons to communicate their intentions could allow a coordination of commitments that otherwise couldn't be guaranteed.

"Quantum mechanics offers the ability to solve the free-rider problem in the absence of a third-party enforcer," wrote Kay-Yut Chen, Tad Hogg, and Raymond Beausoleil in their paper.[8]

QUANTUM VOTING

The same principle could be applied to other sorts of community communication issues, including voting, especially in elections with multiple candidates. You wouldn't need runoffs, since the multiple possible outcomes could be encoded in quantum information.

Here, I think, is a real potential for coping with some of the mathematical problems inherent in today's democratic system of voting. For instance, when three candidates are running for office, the ultimate winner may not reflect the desire of the majority of the voters. Here's how it can work:

In a primary election, Candidate A gets 37 percent of the votes, Candidate B gets 33 percent, and Candidate C gets 30 percent. So candidate A and B get into a runoff. But for most of candidate B's voters, C was the second choice. And for most of Candidate A's voters, C was also the second choice. So if C were running against A alone, C would win. If C were running against B alone, C would win. But in the primary, C finished third so the ultimate winner will be A or B. Since a majority of the voters prefer C to either A or B, the winner is clearly not the electorate's optimal choice. A quantum voting scheme could, by incorporating multiple possibilities in the voting, reach a more "democratic" election result.

It sounds far-fetched, but its mere possibility affirms the potentially dramatic value of invoking quantum weirdness to cope with the complexities of the ordinary world. And it may even be possible that quantum game theory underlies much deeper aspects of nature and of life. In the mushrooming literature on quantum games are papers suggesting that quantum strategies at the molecular level may mimic aspects of evolutionary game-theoretic descriptions of the competition between organisms. In particular, Azhar Iqbal of the University of Hull in England argues that quantum entanglement could influence the interactions of molecules leading to a more stable mix of ingredients than would otherwise occur (in analogy to an evolutionary stable strategy for organisms in an ecosystem). A quantum entanglement "strategy," he suggests, could determine whether a population of molecules can "withstand invasion" from a small number of new molecules (corresponding to mutants in evolutionary biology).[9] If there's anything to this—and it would seem to be far too early to say—then you could imagine something like quantum game theory playing a role in the origin of stable sets of self-replicating molecules—in other words, life itself. (In which case the Code of Nature might turn out to be solvable only with quantum cryptography.)

In any case, quantum game theory offers a new perspective on both games and physics, with implications awaiting further exploration. At the very least, quantum physics and games share one obvious similarity—probability distributions, as with the mixed strategies of games and the mixed realities of quantum mechanics. Life and physics, it seems, are all mixed up. Sorting it all out will require a closer look at the power of probability.

11

Pascal's Wager

Games, probability, information, and ignorance

> Although this may seem a paradox, all exact science is
> dominated by the idea of approximation.
> —Bertrand Russell

As a teenager in 17th-century France, Blaise Pascal seemed destined for mathematical greatness. He wrote a genius-caliber treatise on geometry at age 16 and invented a rudimentary computer to assist the calculations of his tax-collector father. But as an adult, Pascal was seduced by religion, forgoing math to produce a series of philosophical musings assembled (after his death) into a book called *Pensées*. He died at 39, leaving a legacy, in the words of the mathematician E. T. Bell, as "perhaps the greatest might-have-been in history."[1]

Still, Pascal remains a familiar name in today's mathematics textbooks, thanks to a favor he did for a French aristocrat who desired assistance with his gambling habit. What Pascal offered was not religious counseling on the evils of gambling, but mathematical advice on how to win. In his correspondence on this question with Pierre Fermat, Pascal essentially invented probability theory. What's more, out of Pascal's religious ruminations came an idea about probability that was to emerge centuries later as a critical concept in mathematics, with particular implications for game theory.

When it comes to making bets, Pascal observed, it is not enough to know the odds of winning or losing. You need to know what's at stake. You might want to take unfavorable odds if the payoff for winning would be really huge, for example. Or you might consider playing it safe by betting on a sure thing even if the payoff was small. But it wouldn't seem wise to bet on a long shot if the payoff was going to be meager.

Pascal framed this issue in his religious writings, specifically in the context of making a wager about the existence of God. Choosing to believe in God was like making a bet, he said. If you believe in God, and that belief turns out to be wrong, you haven't lost much. But if God does exist, believing wins you an eternity of heavenly happiness. Even if God is a low-probability deity, the payoff is so great (basically, infinite) that He's a good bet anyway. "Let us weigh the gain and the loss in wagering that God is," Pascal wrote. "Let us estimate these two chances. If you gain, you gain all; if you lose, you lose nothing. Wager, then, without hesitation that He is."[2]

Pascal's reasoning may have been theologically simplistic, but it certainly was mathematically intriguing.[3] It illustrated the kind of reasoning that goes into calculating the "mathematical expectation" of an economic decision—you multiply the probability of an outcome by the value of that outcome. The rational choice is the decision that computes to give the highest expected value. Pascal's wager is often cited as the earliest example of a math-based approach to decision theory.

In real life, of course, people don't always make their decisions simply by performing such calculations. And when your best decision depends on what other people are deciding, simple decision theory no longer applies—making the best bets becomes a problem in game theory. (Some experts would say decision theory is just a special case of game theory, in which one player plays the game against nature.) Still, probabilities and expected payoffs remain intertwined with game theory in a profound and complicated way.

For that matter, all of science is intertwined with probability theory in a profound way—it's essential for the entire process of observation, experiment, and measurement, and then comparing those numbers with theory. And probability arises not only in making measurements and testing hypotheses, but also in the very description of physical phenomena, particularly in the realm of statistical physics. In the social sciences, of course, probability theory is also indispensable, as Adolphe Quetelet argued almost two centuries ago. So game theory's intimate relationship with probabilities, I'd wager, is one of the reasons why it finds such widespread applicability in so many different scientific contexts. And no doubt it's this aspect of game theory that has positioned it so strategically as an agent for merging social and physical statistics into a physics of society—something like Asimov's psychohistory or a Code of Nature.

So far, attempts to devise a sociophysics for describing society have mostly been based not on game theory, but on statistical mechanics (as was Asimov's fictional psychohistory). But game theory's mixed strategy/probabilistic formulas exhibit striking similarities to the probability distributions of statistical physics. In fact, the mixed strategies used by game players to achieve a Nash equilibrium *are* probability distributions, precisely like the distributions of molecules in a gas that statistical physics quantifies.

This realization prompts a remarkable conclusion—that, in a certain sense, game theory and statistical mechanics are alter egos. That is to say, they can be expressed using the same mathematical language. To be more precise, you'd have to say that certain versions of game theory share math identical to particular formulations of statistical mechanics, but the deep underlying connection remains. It's just that few people have noticed it.

STATISTICS AND GAMES

If you search the research literature thoroughly, though, you will find several papers from the handful of scientists who have begun to exploit the game theory–statistical physics connection. Among them is David Wolpert, a physicist-mathematician at NASA's Ames Research Center in California.

Wolpert is one of those creative thinkers who refuse to be straitjacketed by normal scientific stereotypes. He pursues his own intuitions and interests along the amorphous edges separating (or connecting) physics, math, computer science, and complexity theory. I first encountered him in the early 1990s while he was exploring the frontiers of interdisciplinary science at the Santa Fe Institute, discussing such issues as the limits of computability and the nature of memory.

In early 2004, Wolpert's name caught my eye when I noticed a paper he posted on the World Wide Web's physics preprint page.[4] His paper showed how to build a bridge between game theory and statistical physics using information theory (providing, incidentally, one of the key inspirations for writing this book). In fact, as Wolpert showed in the paper that attracted my attention to this issue in the first place, a particular approach to statistical mechanics turns out to use math that is equivalent to the math for non-cooperative games.

Wolpert's paper noted that the particles described by statistical physics are trying to minimize their collective energy, like the way people in a game try to reach the Nash equilibrium that maximizes their utility. The mixed strategies used by players to achieve a Nash equilibrium are probability distributions, just like the distribution of energy among particles described by statistical physics.

After reading Wolpert's paper, I wrote him about it and then a few months later discussed it with him at a complexity conference outside Boston where he was presenting some related work. I asked what had motivated him to forge a link between game theory and statistical physics. His answer: rejection.

Wolpert had been working on collective machine learning systems, situations in which individual computers, or robots, or other autonomous devices with their own individual goals could be co-ordinated to achieve an objective for the entire system. The idea is to find a way to establish relationships between the individual "agents" so that their collective behavior would serve the global goal. He noticed similarities in his work to a paper published in *Physical Review Letters* about nanosized computers. So Wolpert sent off one of his papers to that journal.

"The editor actually came back and said 'Well, . . . what you're doing just plain isn't physics,'" Wolpert said. "And I was annoyed." So he started thinking about physics and game theory. After all, a bunch of agents with their own agendas, yet pursuing a common goal, is entirely analogous to players in a game seeking a Nash equilibrium. "And then I said, OK, I'm going to try to take that and completely translate it into a physics system," he recalled.[5]

Games deal with players; physics deals with molecules. So Wolpert worked on the math that would represent a player's strategy like a molecule's state of motion. The mix of all the players' strategies would then be like the combined set of motion states of all the molecules, as ordinarily described by statistical physics. The formulas he came up with would allow you to calculate a good approximation to the actual set of any individual player's strategies in a game, given some limited knowledge about them. You could then do exactly the same sort of calculation for the combined strategies of all the players in a game. Basically, Wolpert showed how the math of statistical physics turns out to be the same as the math for games where players have limited rationality.

"Those topics are fundamentally one and the same," he wrote in his paper. "This identification raises the potential of transferring some of the powerful mathematical techniques that have been developed in the statistical physics community to the analysis of noncooperative game theory."[6]

Wolpert's mathematical machinations were rooted in the idea of "maximum entropy," a principle relating standard statistical physics to information theory, the math designed to quantify the sending and receiving of messages. The maximum entropy (or "maxent") idea was promoted by the maverick physicist Edwin Jaynes in a 1957 paper that was embraced by a number of physicists but ignored by many others. Wolpert, for one, calls Jaynes's work "gloriously beautiful" and thinks that it's just what scientists need in order "to bring game theory into the 21st century."

Jaynes's principle is simultaneously intriguing and frustrating. It seems essentially simple but nevertheless poses tricky complications. It is intimately related to the physical concept of entropy, but is still subtly different. In any event, its explanation requires a brief excursion into the nature of probability theory and information theory, the essential threads tying game theory and statistical physics together.

PROBABILITY AND INFORMATION

For centuries, scientists and mathematicians have argued about the meaning of probability. Even today there exist separate schools of probabilistic thought, generally referred to by the shorthand labels of "objective" and "subjective." But those labels conceal a tangle of subarguments and technical subtleties that make probability theory one of the most contentious and confusing realms of math and science.

In a way, that's a bit surprising, since probability theory really lies at the very foundation of science, playing the central role in the process of analyzing experimental data and testing theories. It's what doing science is all about. You'd think they'd have it all worked out by now. But establishing rules for science is a little like framing a constitution for Iraq. There are different philosophies and approaches to science. The truth is that science (unlike mathematics) is not built on a rock-solid foundation of irreducible rules. Science is like grammar. Grammar arises from regularities that evolve in the way native speakers of a language form their words and string them together. A true grammarian does not tell people how they should speak, but codifies the way that people actually do speak. Science does not emanate from a cookbook that provides recipes for revealing nature's secrets, but from a mix of methods that somehow succeed in rendering nature comprehensible. That's why science is not all experiment, and not all theory, but a complex interplay of both.

Ultimately, though, theory and experiment must mesh if the scientist's picture of nature is to be meaningful and useful. And in most realms of science, you need math to test the mesh. Probability theory is the tool for performing that test. (Different ideas about how to perform the test, then, lead to different conceptions of probability.)

Before Maxwell, probability theory in science was mostly limited to quantifying things like measurement errors. Laplace and others showed the way to estimate how far off your measurement was likely to be from the true value for a particular degree of confidence. Laplace himself applied this approach to measuring the mass of Saturn. He concluded that there was only one chance in 11,000 that the true mass of Saturn would deviate from the then-current measurement by more than 1 percent. (As it turned out, today's best measurement indeed differs from the one in Laplace's day by only 0.6 percent.) Probability theory has developed into an amazingly precise way of making such estimates.

But what does probability itself really mean? If you ask people who ought to know, you'll get different answers. The "objective" school of thought insists that the probability of an event is a property of the event. You observe in what fraction of all cases that event happens and thereby measure its objective probability. The subjective view, on the other hand, argues that probability is a *belief* about how likely something is to happen. Measuring how often something happens gives you a frequency, not a probability, the subjectivists maintain.

There is no point here delving into the debates about the relative merits of these two views. Dozens of books have been devoted to that controversy, which is largely irrelevant to game theory. The fact is that the prevailing view today, among physicists at least, is that the subjectivist approach contains elements that are essential for a sound assessment of scientific data.

Subjective statistics often goes under the label of Bayesian, after Thomas Bayes, the English clergyman who discussed an approach of that nature in a paper published in 1763 (two years after his death). Today a formula known as Bayes' theorem is at the heart of practicing the subjective statistics approach (although that precise theorem was actually worked out by Laplace). In any case, the Bayesian viewpoint today comes in a variety of flavors, and there is much disagreement about how it should be interpreted and applied (perhaps because it is, after all, subjective).

From a practical point of view, though, the math of objective and subjective probability theory does not really differ in any fundamental respect other than its interpretation. It's just that in some cases it seems more convenient, or more appropriate, to use one rather than another, as Jaynes pointed out half a century ago.

INFORMATION AND IGNORANCE

In his 1957 paper,[7] Jaynes championed the subjectivity side of the probability debate. He noted that both views, subjectivist and objectivist, were needed in physics, but that for some types of problems only the subjective approach would do.

He argued that the subjective approach can be useful even when you know nothing about the system you are interested in to begin with. If you are given a box full of particles but know nothing about them—not their mass, not their composition, not their internal structure—there's not much you can say about their behavior. You know the laws of physics, but you don't have any knowledge about the system to apply the laws to. In other words, your ignorance about the behavior of the particles is at a maximum.

Early pioneers of probability theory, such as Jacob Bernoulli and Laplace, said that in such circumstances you must simply assume that all the possibilities are equally likely—until you have some reason to assume otherwise. Well, that helps in doing the calculations, perhaps, but is there any real basis for assuming the probabilities are equal? Except for certain cases where an obvious symmetry is at play (say, a perfectly balanced two-sided coin), Jaynes said, many other assumptions might be equally well justified (or the way he phrased it, any other assumption would be equally arbitrary).[8]

Jaynes saw a way of coping with this situation, though, with the help of the then fairly new theory of information devised by Claude Shannon of Bell Labs. Shannon was interested in quantifying communication, the sending of messages, in a way that would help engineers find ways to communicate more efficiently (he worked for the telephone company, after all). He found math that could quantify information quite nicely if you viewed communication as the reduction of uncertainty. Before communication begins, all messages are possible, so uncertainty is high; as messages are actually received, that uncertainty is reduced.

Shannon's math applied generally to any system of signaling, from Morse Code to smoke signals. But suppose, for example, that all you wanted to do was send someone a single English word (from among all the words in a standard unabridged dictionary, about half a million). If you tell the recipient of the message that the word is in the first half of the dictionary, you've reduced the number of possibilities from 500,000 to 250,000. In other words, you have reduced the uncertainty by half (which so happens to correspond to one bit of information).

Shannon elaborated on this idea to show how all communication could be quantified based on the idea that messages reduce uncertainty. He found a formula for a quantity that measures that uncertainty precisely—the greater the uncertainty, the greater the quantity. Shannon called the quantity entropy, a conscious analogy to the entropy term used by physicists in statistical mechanics and thermodynamics.

Physicists' entropy is a measure of the disorder in a physical system. Suppose you have a chamber containing separate compartments, and you place a zillion molecules of oxygen in the left-side compartment and 4 zillion molecules of nitrogen in the right-side compartment. Then you remove the partition between the compartments. The molecules soon get all mixed up—more disordered—and so the entropy of the system has increased. But something else has happened—you no longer know where the molecules are. Your ignorance of their location has increased just as the entropy has increased. Shannon showed that his formula for entropy in communication—as a measure of ignorance or uncertainty—is precisely the same equation that is used in statistical mechanics to describe the increasing entropy of a collection of particles.

Entropy, in other words, is the same thing as ignorance. Entropy is synonymous with uncertainty. Information theory therefore provides a precise new way of measuring uncertainty in a probability distribution.

So here's a clue about what to do when you know nothing about the probabilities in the system you want to study. Choose a probability distribution that maximizes the entropy! Maximum entropy means maximum ignorance, and if you know nothing, ignorance is by definition at a maximum. Assuming maximum entropy/ignorance, then, is not just an assumption; it's a factual statement about your situation.

Jaynes proposed that this notion of maximum ignorance should be elevated to the status of a basic principle for describing anything scientifically. In his view, statistical mechanics itself just became a system of statistical inference about a system. By taking the maxent approach, you still get all the computational rules that statistical mechanics provides, without the need to assume anything at all about the underlying physics.

In particular, you now can justify the notion that all the possibilities are equally possible. The whole idea is that no possibility (allowed by the laws of physics) gets left out. Everything not explicitly excluded by the information you've got has to be viewed as having some probability of occurring. (In standard statistical mechanics, that feature was simply assumed without evidence—probability distributions were based on the idea that molecules would explore all their possible behaviors.) And if you know nothing, you cannot say that any one possibility is more likely than any other—that would be knowledge.

Of course, if you *know* something about the probabilities, you can factor that in to the probability distribution you use to make your predictions about what's going to happen. But if you know nothing at all, there's only one probability distribution that you can identify for making your bets: *the one that has the maximum entropy, the maximum uncertainty, the maximum ignorance.* It makes sense, after all, because knowing nothing is, in fact, being maximally ignorant.

This is the magic that makes it possible to make a prediction even when knowing nothing about the particles or people you're making the prediction about. Of course, your prediction might not turn out to be right. But it's still the best possible prediction you can make, the likeliest answer you can identify, when you know nothing to begin with.

"The fact that a probability distribution maximizes the entropy subject to certain constraints becomes the essential fact which justifies use of that distribution for inference," Jaynes wrote. "Whether or not the results agree with experiment, they still represent the best estimates that could have been made on the basis of the information available."[9]

But what, exactly, does it mean to "maximize the entropy"? It simply means choosing the probability distribution that would result from adding up all the possibilities permitted by the laws of nature (since you know nothing, you cannot leave out anything that's possible). Here's a simple example. Suppose that you want to predict the average grade for a class of 100 students. All you know are the rules (that is, the laws of nature)—everybody gets a grade, and the grade has to be A, B, C, D, or F (no incompletes allowed). You don't know anything about the caliber of the students or how hard the class is. What is your best prediction of the average grade for the kids in the class? In other words, how do you find a probability distribution for the grades that tells you which grade average is the most probable?

Applying the maxent or maximum ignorance principle, you simply assume that the grades can be distributed in all possible ways—all possible combinations equally likely. For instance, one possible distribution is 100 A's and nothing else. Another would be all F's. There could be 20 students with each grade. You could have 50 C's, 20 B's and 20 D's, 5 A's and 5 F's. All the combinations sum to an ensemble of possibilities that constitutes the probability distribution corresponding to no knowledge—maximum ignorance—about the class and the kids and their grades.

In statistical physics, this sort of thing is called the "canonical ensemble"—the set of possible states for the molecules in a system. Each possible combination is a microstate. Many different possible microstates (distributions of grades) can correspond to the same average (the macrostate).

Don't try to list all the possible combinations; it would take you a very long time. (You're talking something close to 10 to the 70th power.) But you can calculate, or even see intuitively, that the most likely average grade will be C. Of all the possible microstate combinations, many more work out to a C average than to any other grade. There is only one way, for instance, to have a perfect A average—all 100 students getting A's. But you can get a C average in many different ways—100 C's, 50 A's and 50 F's, 20 students getting each of the five grades, and so on.[10]

It's just like flipping pennies, four flips at a time, with the grade corresponding to the number of heads that turn up (0 = F, 4 = A). In 100 trials, many combinations give an average of 2, but only a few will give an average of 0 or 4. So your prediction, based on knowing nothing, will be an average grade of C.

BACK TO THE GAME

In game theory, a player's mixed strategy is also a probability distribution, much like grades or penny flips. Game theory is all about how to figure out what each player's best mixed strategy would be (for maximizing utility, or the payoff, of the game). In a multiplayer game, there is at least one mix of all players' mixed strategies for which no one player could do any better by changing strategies—the Nash equilibrium, game theory's most important foundational principle.

But Nash's foundation of modern game theory has its cracks. While it's true that, as Nash showed, all games (with certain qualifications) have at least one Nash equilibrium, many games can have more than one. In those cases, game theory cannot predict which equilibrium point will be reached—you can't say what sets of mixed strategies the players will actually adopt in a real-world situation. And even if there is only one Nash equilibrium in a complicated game, it is typically beyond the capability of a committee of supercomputers to calculate what all the players' mixed strategies would have to be.

In turn, that crack is exacerbated by a weakness in the cardinal assumption underlying traditional game theory—that the players are rational payoff maximizers with access to all the necessary information to calculate their payoffs. In a world where most people can't calculate the sales tax on a cheeseburger, that's a tall order. In real life, people are not "perfectly rational," capable of figuring out the best money-maximizing strategy for any strategy combination used by all the other competitors. So game theory appears to assume that each player can do what supercomputers can't. And in fact, almost everybody recognizes that such total rationality is unachievable. Modern approaches to game theory often assume, therefore, that rationality is limited or "bounded."

Game theorists have devised various ways to deal with these limitations on Nash's original math. An enormous amount of research, of the highest caliber, has modified and elaborated game theory's original formulations into a system that corrects many of these initial "flaws." Much work has been done on understanding the limits of rationality, for instance. Nevertheless, many game theorists often cling to the idea that "solving a game" means finding an equilibrium—an outcome where all players achieve their maximum utility. Instead of thinking about what will happen when the players actually play a game, game theorists have been asking what the individual players should do to maximize their payoff.

When I visited Wolpert at NASA Ames, a year after our conversation in Boston, he pointed out that the search for equilibrium amounts to viewing a game from the inside, from the viewpoint of one of the participants, instead of from the vantage point of an external scientist assessing the whole system. From the inside, there may be an optimal solution, but a scientist on the outside looking in should merely be predicting what will happen (not trying to win the game). If you look at it that way, you know you can never be sure how a game will end up. A science of game theory should therefore not be seeking a single answer, but a probability distribution of answers from which to make the best possible prediction of how the game will turn out, Wolpert insists. "It's going to be the case that whenever you are given partial information about a system, what must pop out at the other end is a distribution over possibilities, not a single answer."[11]

In other words, scientists in the past were not really thinking about the game players as particles, at least not in the right way. If you think about it, you realize that no physicist computing the thermodynamic properties of a gas worries about what an individual molecule is doing. The idea is to figure out the bulk features of the whole collection of molecules. You can't know what each molecule is up to, but you can calculate, statistically, the macroscopic behavior of all the molecules combined. The parallel between games and gases should be clear. Statistical physicists studying gases don't know what individual molecules are doing, and game theorists don't know what individual players are thinking. But physicists do know how collections of molecules are likely to behave—statistically—and can make good predictions about the bulk properties of a gas. Similarly, game theorists ought to be able to make statistical predictions about what will happen in a game.

This is, as Wolpert repeatedly emphasizes, the way science usually works. Scientists have limited information about the systems they are studying and try to make the best prediction possible given the information they have. And just as a player in a game has incomplete information about all the game's possible strategy combinations, the scientist studying the game has incomplete information about what the player knows and how the player thinks (remember that different individuals play games in different ways).

All sciences face this sort of problem—knowing something about a system and then, based on that limited knowledge, trying to predict what's going to happen, Wolpert pointed out. "So how does science go about answering these questions? In every single scientific field of endeavor, what will come out of such an exercise is a probability distribution."[12]

From this point of view, another sort of mixed strategy enters game theory. It's not just that the player has a mixed strategy, a probability distribution of possible moves from which to choose. The scientist describing the game also has a kind of "mixed strategy" of possible predictions about how the game will turn out.

"When you think about it, it's obvious," Wolpert said. "If I give you a game of real human beings, no, you're not going to always have the same outcome. You're going to have more than one possible outcome. . . . It's not going to be the case they are always going to come up with the exact same set of mixed strategies. There's going to be a distribution over their mixed strategies, just like in any other scientific scenario."

This is clearly taking game theory to another level. While each player has a mixed strategy, a probability distribution of pure strategies, the scientist describing the game should compute a probability distribution of all the players' mixed strategies. And how do you find those probability distributions of mixed strategies? By maximizing your ignorance, of course. If you want to treat game theory as though the people were particles, the best approach is to assume a probability distribution for their strategies that maximizes the uncertainty (or the entropy, in information theory terms). Using this approach, you don't need to *assume* that the players in a game have limits on their rationality; such limits naturally appear in the formulas that information theory provides. Given a probability distribution of possible outcomes for the game, then, you can choose which outcome to bet on using the principles of decision theory.

"When you need a prediction, a probability distribution won't do," said Wolpert. "You have to decide to fire the missile or don't fire; turn left or right." The underlying axioms for the mathematical basis for making such a decision were worked out in the 1950s by Leonard Savage[13] in some detail, but they boil down to something like Pascal's Wager. If you have a probability distribution of possible outcomes, but don't know enough to distill the possibilities down to a single prediction, you need to consider how much you have to lose (or to gain) if your decision is wrong (or right).

"If you predict X, but the truth is Y, how much are you hurt?

Or conversely, how much do you benefit?" Wolpert explained. "Certain kinds of mispredictions aren't going to hurt you very much, depending on what the truth is. But in other instances . . . your prediction of truth might cause all sorts of problems—you've now launched World War III."

Decision theory dictates that you should make the prediction that minimizes your expected loss ("expected" signifying that the relative probabilities of the choices are taken into account—you average the magnitudes of loss over all the possibilities). Consequently, Wolpert observes, different individual observers would make different predictions about the outcome of a game, even if the probability distribution of possible outcomes is the same, because some people would have more to lose than others for certain incorrect predictions.

"In other words, for the exact same game, your decision as the external person making the prediction is going to vary depending on your loss function," he says. That means the best prediction about the outcome isn't some equilibrium point established within the game, but rather depends on "the person external to the game who's making the prediction about what's going to come out of it." And so sometimes the likeliest outcome of a game will *not* be a Nash equilibrium.

But why not, if a Nash equilibrium represents the stable outcome where nobody has an incentive to change? It seems like people would keep changing their strategy until they had no incentive not to. But when game theory is cast in the information-theoretic equations of maximum entropy, the answer becomes clear. A term in the equations signifies the cost of computing the best strategy, and in a complicated game that cost is likely to be too high. In other words, a player attempting to achieve a maximum payoff must factor in the cost of computing what it takes to get that payoff. The player's utility is not just the expected payoff, but the expected payoff minus the cost of computing it.

What's more, individual differences can influence the calculations. The math of the maximum ignorance approach (that is, maximizing the uncertainty) contains another term, one that can be interpreted as a player's temperature. Temperature relates ignorance (or uncertainty) to the cost of computing a strategy—more uncertainty about what to do means a higher cost of figuring out what to do. A low temperature signifies a player who focuses on finding the best strategy without regard to the cost of computing it; a higher-temperature player will explore more of the strategy possibilities.

"So what that means," Wolpert explained, "is that it is literally true that somebody who is purely rational, who always does the best possible thing, is cold—they are frozen. Whereas somebody who is doing all kinds of things all over the map, exploring, trying all kinds of possibilities, they are quite literally hot. That just falls out of the math. It's not even a metaphor; it's what it actually amounts to."[14]

Temperature, in other words, represents a quantification of irrationality. In a gas, higher temperatures mean there's a higher chance that the molecules are not in the arrangement that minimizes their energy. With game players, higher temperature means a greater chance that they won't be maximizing their payoff.

"The analogy is that you have some probability of being in a nonpurely rational state," Wolpert said. "It's the exact same thing. Lowering energy is raising utility." You are still likely to play strategies that would increase your payoff, but just how much more likely depends on your temperature.[15]

Boiled down to the key point, the maximum entropy math tells you that game players will have limited rationality—it's not something that you have to assume. It arises naturally from adopting the viewpoint of somebody looking from outside the game instead of being inside the game.

"That is crucial," Wolpert stressed. "Game theory has always had probability theory inside of it, because people play mixed strategies, but game theory has never actually applied probability theory to the game as a whole. That is the huge hole in conventional game theory."

Ultimately, the idea of a player's temperature should allow better predictions of how real players will play real games. In the probability distribution of grades in a class, the maximum entropy approach says all grade distributions are possible. But if you know something about the students—maybe all are honors students who've never scored below a B—you can adjust the probability distribution by adding that information into the equations. If you know something about a player's temperature—the propensity to explore different possible strategies—you can add *that* information into the equations to improve your probability distribution. With collaborators at Berkeley and Purdue, Wolpert is beginning to test that idea on real people—or at least, college students.

"We've just run through some experiments on undergrads where we're actually looking at their temperatures, in a set of repeated games—voting games in this case—and seeing things like how does their temperature change with time. Do they actually get more rational or less rational? What are the correlations between different individuals' temperatures? Do I get more rational as you get less rational?"

If, for instance, one player is always playing the exact same move, that makes it easier for opponents to learn what to expect. "That suggests intuitively that if you drop your temperature, mine will go up," Wolpert said. "So in these experiments our intention is to actually look for those kinds of effects."

VISIONS OF PSYCHOHISTORY

Such experiments, it seemed to me, would add to the knowledge that behavioral game theorists and experimental economists had been accumulating (including inputs from psychology and neuroeconomics) about human behavior. It sounded like Wolpert was saying that all this knowledge could be fed into the probability distribution formulas to improve game theory's predictive power. But before I could ask about what was really on my mind, he launched into an elaboration that took me precisely where I wanted to go.

"Let's say that you know something from psychology, and you've gotten some results from experiments," he said. "Then you actually have other stuff that goes in here [the equations] besides the knowledge that all human beings have temperatures. You also know something about their degree of being risk averse, and this, that, or the other. . . . You are not just a temperature; there are other aspects to you."

Adding such knowledge about real people into the equations reduces the ignorance that went into the original probability distribution. So instead of predictions based on all possible mixed strategies, you'll get predictions that better reflect real people. "It's a way of actually integrating game theory with psychology, formally," Wolpert said. "You would have . . . quantification of individual human beings' behavior integrated with an actual mathematical structure that deals with incentives and utility functions and payoffs."

Wolpert began talking about probability distributions of future states of the stock market and then, almost as an aside, disclosed a much grander vision. "This actually is a way of trying to get a mathematics of psychohistory in Isaac Asimov's sense," Wolpert said. "In other words, this is potentially—it's not been done—this is potentially the physics of human behavior."[16]

Just as I had suspected. The suggestive similarities between Asimov's psychohistory and game theory's behavioral science do, in fact, reflect a common underlying mathematics. It's the math that merges game theory with statistical physics. So in pondering what Wolpert said, it occurred to me that there's a better way to refer to the science of human behavior than psychohistory or sociophysics or Code of Nature. It should be called Game Physics.

Alas, "game physics" is already taken—it's a term used by computer programmers to describe how objects move and bounce around in computerized video games. But it captures the idea of psychohistory or sociophysics pretty well. Game theory combined with statistical physics, the physics of games, is the science of society.

Epilogue

Let the physical basis of a social economy be given—
or, to take a broader view of the matter, of a society.
According to all tradition and experience human beings have a characteristic way of adjusting themselves
to such a background. This consists of not setting up
one rigid system of apportionment . . . but rather a
variety of alternatives, which will probably all express
some general principles but nevertheless differ among
themselves in many particular respects. This system
. . . describes the 'established order of society' or 'accepted standard of behavior.'

—Von Neumann and Morgenstern,
Theory of Games and Economic Behavior

Despite its title, the science fiction cult classic *Ender's Game* isn't
really about game theory, at least not explicitly. But implicitly it is.
It's all about choosing strategies to achieve goals—about adults
plotting methods for manipulating young Ender Wiggin, Ender
choosing among maneuvers to win on a simulated battlefield, and
Ender's siblings' devising tactics for influencing public opinion.
And two passages from Orson Scott Card's novel sound like they
could have been quoted from a game theory textbook, as they
illustrate aspects of human nature that game theory has evolved to
explain. Ender's brother Peter, for instance, epitomizes the selfish
rational agent of game theory's original naive formulation:

Peter could delay any desire as long as he needed to; he could conceal any emotion. And so Valentine knew that he would never hurt her in a fit of rage. He would only do it if the advantages outweighed the risks. . . . He always, always acted out of intelligent self-interest.[1]

Ender himself represents the social actor who plays games with a combination of calculation and intuition, more in line with the notion of game theory embraced by today's behavioral game theorists:

"Every time, I've won because I could understand the way my enemy thought. From what they did. I could tell what they thought I was doing, how they wanted the battle to take shape. And I played off of that. I'm very good at that. Understanding how other people think."[2]

That is, after all, what the modern science of game theory is all about—understanding how other people think. And consequently being able to figure out what they will choose to do. It is also what Isaac Asimov's fictional psychohistory was all about, and what the centuries-long quest by social scientists has been all about—discerning the drumbeat to which society dances. Discovering the Code of Nature.

The modern search for a Code of Nature began in the century following Newton's *Principia*, which established the laws of motion and gravity as the rational underpinning of physical reality. Philosophers and political economists such as David Hume and Adam Smith sought a science of human behavior in the image of Newtonian physics, pursuing the dream that people could be described as precisely as planets. That dream persisted through the 19th century into the 20th, from Adolphe Quetelet's desire to describe society with numbers to Sigmund Freud's quest for a deterministic physics of the brain. Along the way, though, the physics model on which the dream was based itself changed, morphing from the rigid determinism of Newton into the statistical descriptions of Maxwell—the same sorts of statistics used, by Quetelet and his followers, to quantify society. By the end of the 20th century, the quest for a Code of Nature was taken up by physicists who wanted to use those statistics to bring the sciences of society and the natural world back together. Because after all, physics—just ask any physicist—is the science of everything.

PHYSICS AND EVERYTHING

Historically, the physicist's notion of everything has been a bit limited, though. For most of the past three centuries, physics concerned itself mostly with matter and the forces guiding its motion; eventually, the study of matter in motion incorporated energy and its transformations. In the century just gone by, Einstein added cosmic time and space to the mix. He even simplified reality's recipe by combining matter with energy and space with time. Through the 20th-century physicist's eyes, then, "everything" comprised mass-energy and space-time.

Toward the end of that century, a number of physicists began to realize that one ingredient was missing. Awakened by the metaphorical power of the digital computer, astute observers realized that information was the glue connecting the outside world to its scientific description. From the second law of thermodynamics to the weirdness of quantum mechanics to the murky milieu of a black hole's interior, physicists found information to be an indispensable element in codifying and quantifying their understanding of nature.

Information opened physicists' eyes to the rest of reality. Information encompassed biology. Biology included people. People created a new universe of realities for physics to contemplate—vast networks of economic, social, and cultural systems and institutions. So physicists began applying their favorite all-purpose tool—statistical mechanics—to everything from the stock market to flu epidemics. It was all very much in the spirit of Isaac Asimov's fictional mathematician, Hari Seldon, who adopted the principles of statistical mechanics to forecast the future. By the dawn of the 21st century, real-life physicists were trying to do almost exactly the same thing that Seldon had done, using statistical mechanics to build mathematical models of society for the purpose of making predictions.

From its beginnings, game theory had expressed similar ambitions. Von Neumann and Morgenstern focused on economics, but clearly viewed economics as simply one (albeit a major) example of social science in general. They believed that their theory of games was a first step toward a mathematical representation of collective behavior, indeed a Code of Nature (their terms were "standard of behavior" or "order of society").

A few years later, John Nash took a second major step toward a mathematics of society by introducing the Nash equilibrium into game theory's arsenal of ideas. If all the competitors in a game pursue their self-interest—attempting to maximize their expected payoff—there is always some combination of strategies that will produce the best deal that everybody can get (given that everybody plays their best). The existence of a Nash equilibrium in any game implied that societies could be stable—nobody having incentive to change their behavior, as any deviation would lower their payoff if everybody else continued to play the same way.

In both von Neumann's and Nash's math, the essential feature was the need for "mixed strategies" to achieve the maximum payoff. Only rarely is one single "pure" strategy consistently your best bet. Your best strategy is typically to choose from among a range of possible choices, with specified probabilities for each choice.

This idea of a mixed strategy is a recurring theme in game theory and its applications to various aspects of life and society. In evolution, nature plays a mixed strategy, generating complex ecosystems containing a wide range of species. The human race plays a mixed strategy, comprising cooperators, competitors, and punishers. Planet Earth's populations represent a mixed strategy of cultures, from the stingy and solitary Machiguenga in Peru to the generous and gregarious Orma in Kenya. Even in the physical realm, quantum physics shows that reality itself is a mixed strategy at the subatomic level, a feature that game theorists may be able to exploit to solve their thorniest dilemmas.

Such a mixture of choices, with specific probabilities of each, is known in mathspeak as a probability distribution. And probability distributions, it just so happens, are what statistical physics deals with as well. Asimov's basis for psychohistory was applying the laws of probability to large numbers of individual humans to forecast collective human behavior, just as statistical physicists calculate probability distributions of large numbers of molecules to predict the properties of a gas or the course of chemical reactions. Like matter and energy, or space and time, game theory and physics are different sides of a coin. As Pat Benatar would say, they belong together. It's a neat, tight fit, and it's a mystery why it took so long for game theory and physics to mutually realize this underlying relationship.

SEPARATED AT BIRTH

Of course, game theory was conceived with fertilization from physical science, as both von Neumann and Nash applied reasoning rooted in statistical physics. Von Neumann referred to the usefulness of statistics in describing large numbers of interacting agents in an economy. Nash alluded to the statistical interactions of reacting molecules in his derivation of the Nash equilibrium. Nash, after all, studied chemical engineering and chemistry at Carnegie Tech before becoming a math major, and his dissertation at Princeton drew on the chemical concept of "mass action" in explaining the Nash equilibrium. Mass action refers to the way that amounts of reacting chemicals determine the reaction's equilibrium condition, a process described by the statistical mechanics of molecular energies. Borrowing the physical concept of equilibrium in chemical systems of molecules, Nash derived an analogous concept of equilibrium in social systems composed of people. Nash's math was about people, but it was based on molecules, and that math embodies the unification of game theory and social science with physics. The seed of the physics-society link resided within Nash's beautiful mind.

That seed has sprouted and grown in unexpected ways, and its fruits are multiplying, feeding progress in a vast range of sciences, from economics, psychology, and sociology to evolutionary biology, anthropology, and neuroscience. Game theory provides the common mathematical language for unifying these sciences, sciences that represent the puzzle pieces that fit together to generate life, mind, and culture—the totality of collective human behavior. The fact that game theory math can also be translated into the mathematics of the physical sciences argues that it is the key to unlocking the real theory of everything, the science that unifies physics with life.

After all, both physical and living systems seek stability, or equilibrium. If you want to predict the way a chemical reaction will proceed or how people will behave, and how the future will evolve, you need to know how to compute an equilibrium. Game theory shows why reaching an equilibrium point requires mixed strategies—and how this need for mixed strategies drives the creation of complexity. In other words, evolution. Game theory describes the evolutionary process that produces mixtures of different species, mixtures of different types of people, mixtures of different strategies that people employ, mixtures of different cultures that arise in the mixture of environments found around the planet.

Game theory describes the evolutionary process that produces complex networks. The brains that choose from a mix of strategies are networks of nerve cells; the societies that exhibit a mixture of cultures are networks of brains. Put it all together, and you get a framework for quantifying nature that really does encompass everything, a framework merging the game theory of the life and social sciences with the statistical physics describing the material world.

Game theory is not, however, the same as the popular "Theory of Everything" that theoretical physicists have long sought. That quest is merely for the equations describing all of nature's basic particles and forces, the math describing the building blocks. Once you know how the pieces of atoms are put together, this view holds, you don't need to worry about everything else. Game theory, though, is precisely about everything else. It's about the realm of life that builds itself upon the universe's physical foundation. It's about how people carve civilization out of that jungle, and it's about the rules of conduct, the established order of society, the "Code of Nature" that results.

DANGER

There has always been a danger in seeking a Code of Nature—a risk that it would be regarded as a dogmatic deterministic view of human behavior, denying the freedom of the human spirit. Some people react very negatively to that sort of thing. The idea that a Code of Nature is inscribed into human genes, advanced in the 1970s under the label sociobiology, evoked a vitriolic response demonstrating how invective often overwhelms intellect. Sociobiology's intellectual descendant, evolutionary psychology, has produced a more elaborate web of evolution-based explanations for human behavior, but its implied prediction of hardwired brains that play pure strategies doesn't mesh well with the findings of modern neurobiology and behavioral anthropology.

Game theory, on the other hand, offers a possible rapprochement between the advocates of genetic power and the defenders of human freedom. Game theory pursues a different sort of path toward the Code of Nature. It acknowledges the power of evolution—in fact, it helps to explain evolution's ability to generate life's complexities. But game theory also explains why the belief that human nature is rooted in biology, while trivially true, is far from the whole story. Game theory poses no universal gene-controlled determinant of human social behavior, but rather requires, as Nash's math showed, a mixed strategy. It demands that people make choices from multiple possible behaviors.

Game theory's potential scientific power is so great, I think, because it is so intellectually commodious—not narrow and confining, but capable of accommodating many seeming contradictions. That's why it can offer an explanatory structure for all the diversity in the world—a mélange of individual behaviors and personalities, the wide assortment of human cultures, the never-ending list of biological species. Game theory encompasses the coexistence of selfishness and sympathy, competition and cooperation, war and peace. Game theory explains the interplay of genes and environment, heredity and culture. Game theory connects simplicity to complexity by reconciling the tension between evolutionary change and stability. Game theory ties the choices of individual people to the collective social behavior of the human race. Game theory bridges the sciences of mind and mindless matter.

Game theory is about putting it all together. It offers a mathematical recipe for making sense of what seems to be a hopelessly messy world, providing a tangible sign that the Code of Nature is not a meaningless or impossible goal for scientists to pursue. And regardless what anyone thinks about the prospects for ultimate success, scientists are certainly pursuing that goal.

"We want to understand human nature," says Joshua Greene, a neuroscientist and philosopher at Princeton. "That, I think, is a goal in and of itself."[3]

Success may still be a long way off. But somewhere in the vision of Asimov's psychohistory lies an undoubtable truth—that all the world's multiple networks, personal and social, interact in multiple ways to generate a single future. From people to cities, corporations to governments, all of the elements of society must ultimately mesh. What appears to be the madness of crowds *must* have a method, and game theory's successes suggest that it's a method that science can discover.

"The idea is really to have, in the end, a seamless understanding of the universe, from the most basic physical elements, the chemistry, the biochemistry, the neurobiology, to individual human behavior, to macroeconomic behavior—the whole gamut seamlessly integrated," says Greene. "Not in my lifetime, though."

Appendix

Calculating a Nash Equilibrium

Consider the simple game discussed in Chapter 2, where Alice and Bob compete to see how much of a debt to Alice that Bob will have to pay back. This is a zero-sum game; Alice wins exactly what Bob loses, and vice versa. The payoffs in the game matrix are the amounts Bob pays to Alice, so Bob's "payoff" in each case is the negative value of the number indicated.

		Bob	
		Bus	Walk
Alice	Bus	3	6
	Walk	5	4

To calculate the Nash equilibrium, you must find the mixed strategies for each player that yield the best expected payoff when the other player is also choosing the best possible mixed strategy. In this example, Alice chooses Bus with probability p, and Walk with probability $1 - p$ (since the probabilities must add up to 1). Bob chooses Bus with probability q and Walk with probability $1 - q$.

Alice can calculate her "expected payoff" for choosing Bus or Walk as follows. Her expected payoff from Bus will be the sum of:

Her payoff from Bus when Bob plays Bus, multiplied by the probability that Bob will play Bus, or 3 times q
   plus
Her payoff from Bus when Bob plays Walk times the probability that Bob plays Walk, or 6 times $(1 - q)$

Her expected payoff from Walk is the sum of:

Her payoff from Walk when Bob plays Bus times the probability that Bob plays Bus, or 5 times q
   plus
Her payoff from Walk when Bob plays Walk times the probability that Bob plays Walk, or 4 times $(1 - q)$

Summarizing,
   Alice's expected payoff for Bus $= 3q + 6(1 - q)$
   Alice's expected payoff for Walk $= 5q + 4(1 - q)$

Applying similar reasoning to calculating Bob's expected payoffs yields:

   Bob expected payoff for Bus $= -3p + -5(1 - p)$
   Bob expected payoff for Walk $= -6p + -4(1 - p)$

Now, Alice's total expected payoff for the game will be her probability of choosing Bus times her Bus expected payoff, plus her probability of choosing Walk times her Walk expected payoff. Similarly for Bob. To achieve a Nash equilibrium, their probabilities for the two choices must be such that neither would gain any advantage by changing those probabilities. In other words, the expected payoff for each choice (Bus or Walk) must be equal. (If the expected payoff was greater for one than the other, then it would be better to play that choice more often, that is, increasing the probability of playing it.)

For Bob, his strategy should not change if

$$-3p + -5(1 - p) = -6p + -4(1 - p)$$

Applying some elementary algebra skills, that equation can be recast as:

$$-3p - 5 + 5p = -6p - 4 + 4p$$

or

$$2p = 1 - 2p$$

so

$$4p = 1$$

Which, solving for p, shows that Alice's optimal probability for playing Bus is

$$p = 1/4$$

So Alice should choose Bus one time out of 4, and Walk 3 times out of 4.

Now, Alice will not want to change strategies when

$$3q + 6(1 - q) = 5q + 4(1 - q)$$

Which, solving for q, gives Bob's optimal probability for choosing Bus:

$$3q + 6 - 6q = 5q + 4 - 4q$$

$$6 = 4q + 4$$

$$2 = 4q$$

$$q = 1/2$$

So Bob should choose Bus half the time and Walk half the time.

Now let's say Alice and Bob decide to play the hawk-dove game, in which the payoff structure is a little more complicated because what one player wins does not necessarily equal what the other player loses. In this game matrix, the first number in the box gives Alice's payoff; the second number gives Bob's payoff.

		Bob	
		Hawk	Dove
	Hawk	-2, -2	2, 0
Alice			
	Dove	0, 2	1, 1

Alice plays hawk with probability p and dove with probability $1 - p$; Bob plays hawk with probability q and dove with probability $1 - q$. Alice's expected payoff from playing hawk is $-2q + 2(1 - q)$. Her expected payoff from dove is $0q + 1(1 - q)$. Bob's expected payoff from hawk is $-2p + 2(1 - p)$; his expected payoff from dove is $0p + 1(1 - p)$.

Bob will not want to change strategies when

$$-2p + 2(1 - p) = 0p + 1(1 - p)$$

$$2 = 1 + 3p$$

$$3p = 1$$

$$p = 1/3$$

So p, Alice's probability of playing hawk, is $1/3$.

Alice will not want to change strategies if

$$-2q + 2(1 - q) = 0q + 1(1 - q)$$

$$4q - 2 = q - 1$$

$$3q = 1$$

$$q = 1/3$$

So q, Bob's probability of playing hawk, is also 1/3. Consequently the Nash equilibrium in this payoff structure is to play hawk one-third of the time and dove two-thirds of the time.

Further Reading

There are dozens and dozens of books on game theory, of which a handful stand out as indispensable to grasping the theory's essential features. Those that I found most useful and illuminating:

Camerer, Colin. *Behavioral Game Theory*. Princeton, N.J.: Princeton University Press, 2003.

Gintis, Herbert. *Game Theory Evolving*. Princeton, N.J.: Princeton University Press, 2000.

Kuhn, Harold W. and Sylvia Nasar, eds. *The Essential John Nash*. Princeton, N.J.: Princeton University Press, 2002.

Luce, R. Duncan and Howard Raiffa. *Games and Decisions*. New York: John Wiley & Sons, 1957.

Williams, J.D. *The Compleat Strategyst: Being a Primer on the Theory of Games of Strategy*. New York: McGraw-Hill, 1954.

Von Neumann, John and Oskar Morgenstern. *Theory of Games and Economic Behavior. Sixtieth-anniversary Edition*. Princeton, N.J.: Princeton University Press, 2004.

Two other readable books were very helpful:

Davis, Morton D. *Game Theory: A Nontechnical Introduction*. Mineola, NY: Dover, 1997 (1983).

Poundstone, William. *Prisoner's Dilemma*. New York: Anchor Books, 1992.

For the rich and complex historical context of the social sciences into which game theory fits, an excellent guide is:

Smith, Roger. *The Norton History of the Human Sciences*. New York: W.W. Norton, 1997.

And for a comprehensive account of attempts to apply physics to the social sciences:

Ball, Philip. *Critical Mass: How One Thing Leads to Another*. New York: Farrar, Straus and Giroux, 2004.

A few additional books and articles of relevance are listed here; many others addressing specific points are mentioned in the notes.

Books

Harman, P.M. *The Natural Philosophy of James Clerk Maxwell*. Cambridge: Cambridge University Press, 1998.

Henrich, Joseph, et al., eds. *Foundations of Human Sociality: Economic Experiments and Ethnographic Evidence from Fifteen Small-Scale Societies*. New York: Oxford University Press, 2004.

Macrae, Norman. *John von Neumann*. New York: Pantheon Books, 1991.

Nasar, Sylvia. *A Beautiful Mind*. New York: Simon & Schuster, 1998.

Watts, Duncan J. *Six Degrees*. New York: W.W. Norton, 2003.

Articles

Ashraf, Nava, Colin F. Camerer, and George Loewenstein. "Adam Smith, Behavioral Economist." *Journal of Economic Perspectives,* 19 (Summer 2005): 131–145.

Ball, Philip. "The Physical Modelling of Society: A Historical Perspective." *Physica A,* 314 (2002): 1–14.

Holt, Charles and Alvin Roth. "The Nash Equilibrium: A Perspective." *Proceedings of the National Academy of Sciences USA,* 101 (March 23, 2004): 3999–4002.

Morgenstern, Oskar. "Game Theory." *Dictionary of the History of Ideas.* Available online at *http://etext.virginia.edu/DicHist/ dict.html.*

Myerson, Roger. "Nash Equilibrium and the History of Economic Theory." 1999. Available online at *http://home.uchicago.edu/ ~rmyerson/research/jelnash.pdf.*

After the manuscript for this book was completed, a new review article appeared exploring the game theory-statistical mechanics relationship in depth:

Szabó, György and Gábor Fáth. "Evolutionary Games on Graphs," *http://arxiv.org/abs/cond-mat/0607344,* July 13, 2006.

Notes

INTRODUCTION

1. Isaac Asimov, *Foundation and Earth,* Doubleday, Garden City, N.Y., 1986, p. 247.

2. Isaac Asimov, *Foundation's Edge*, Ballantine Books, New York, 1983 (1982), p. xi.

3. Herbert Gintis, *Game Theory Evolving,* Princeton University Press, Princeton, N.J., 2000, pp. xxiv–xiv.

4. Samuel Bowles, telephone interview, September 11, 2003.

5. Read Montague, interview in Houston, Tex., June 24, 2003.

6. Gintis, *Game Theory Evolving,* p. xxiii.

7. Asimov, *Foundation and Earth,* p. 132.

8. Stephen Wolfram, in his controversial book *A New Kind of Science,* also claims to show a network-related way of explaining quantum physics—and everything else in the universe. If he is right, game theory may someday have something to say about the universe as well.

SMITH'S HAND

1. Jacob Bronowski and Bruce Mazlish, *The Western Intellectual Tradition,* Harper & Row, New York, 1960, p. 353.

2. David Hume, *A Treatise of Human Nature*, available online at *http://etext.library.adelaide.edu.au/h/hume/david/h92t/introduction.html.*

3. James Anson Farrer, *Adam Smith*, Sampson, Low, Marston, Searle and Rivington, 1881, p. 2. Available online at *http://socserv2.socsci.mcmaster.ca/~econ/ugcm/3ll3/smith/farrer.html.*

4. Adam Smith, *The Wealth of Nations*, Bantam, New York, 2003 (1776).

5. Ibid., pp. 23–24, 572.

6. Alan Krueger, Introduction, *The Wealth of Nations,* Bantam, New York, 2003, p. xviii.

7. Ibid., p. xxiii.

8. Thomas Edward Cliffe Leslie, "The Political Economy of Adam Smith," *The Fortnightly Review*, November 1, 1870. Available online at *http:// etext.lib.virginia.edu/modeng/modengS.browse.html*

9. "Code of Nature" was most unfortunately used in the title of a work by a French communist named Morelly. He had truly wacko ideas. I don't mean to pick on communists—they've had a bad enough time in recent years—but this guy really was off the charts. For one thing, he insisted that everybody had to get married whether they wanted to or not. And you had to turn 30 years old before you would be allowed to pursue an academic profession if you so desired, provided you were judged worthy.

10. Henry Maine, *Ancient Law*, 1861. Available online at *http://www.yale.edu/ lawweb/avalon/econ/maineaco.htm*. Maine notes that "Jus Gentium was, in actual fact, the sum of the common ingredients in the customs of the old Italian tribes, for they were all the nations whom the Romans had the means of observing, and who sent successive swarms of immigrants to Roman soil. Whenever a particular usage was seen to be practiced by a large number of separate races in common it was set down as part of the Law common to all Nations, or Jus Gentium."

11. Dugald Stewart, "Account of the Life and Writings of Adam Smith LL.D.," *Transactions of the Royal Society of Edinburgh*, 1793. Available online at *http:// socserv2.socsci.mcmaster.ca/~econ/ugcm/3ll3/smith/dugald*.

12. Cliffe Leslie, "Political Economy," pp. 2, 11.

13. Roger Smith, *The Norton History of the Human Sciences,* W.W. Norton, New York, 1997, p. 303.

14. Colin Camerer, interview in Pasadena, Calif., March 12, 2004.

15. Nava Ashraf, Colin F. Camerer, and George Loewenstein, "Adam Smith, Behavioral Economist," *Journal of Economic Perspectives,* 19 (Summer 2005): 132.

16. Stephen Jay Gould, *The Structure of Evolutionary Theory,* Harvard University Press, Cambridge, Mass., 2002, pp. 122–123.

17. Charles Darwin, *The Origin of Species*, The Modern Library, New York, 1998, p. 148.

18. Another interesting refutation of Paley comes from Stephen Wolfram, whose book *A New Kind of Science* generated an enormous media blitz in 2002. Wolfram makes the point that a Swiss watch—Paley's example of complexity—is actually quite a simple, regular, predictable device. You need a designer, Wolfram said, not to produce complexity, but to ensure simplicity.

A watch, after all, exhibits nothing like the complexity of life, Wolfram pointed out. Keeping time requires, above all else, absolutely regular motion to guarantee near-perfect predictability. Complexity introduces deviations from regular motion, rendering a clock worthless. And as Wolfram demonstrates throughout his book, nature—left to its own devices—produces complexity with wild abandon. In the biological world, such complexity is messy and unpredictable, and for that sort of thing you need no designer at all, just simple rules governing how a system evolves over time (and some of those rules might be provided by game theory). A Swiss watch, on the other hand, does not evolve over time—it just tells time. And it has no offspring, only springs.

19. Gould, *Evolutionary Theory*, p. 124.

20. Of course, had Einstein introduced relativity in a book, instead of writing scientific papers, the 20th century would have had a similar masterpiece.

VON NEUMANN'S GAMES

1. Maria Joao Cardoso De Pina Cabral, "John von Neumann's Contribution to Economic Science," *International Social Science Review*, Fall–Winter 2004. Available online at *http://www.findarticles.com/p/articles/mi_m0IMR/is_3-4_79*.

2. Jeremy Bentham, *An Introduction to the Principles of Morals and Legislation*, Clarendon Press, Oxford, 1907 (1789), Chapters I, III. While written in 1780 and distributed privately, it wasn't published until 1789.

3. Jeremy Bentham, *A Fragment on Government*, London, 1776, Preface. Available online at *http://www.ecn.bris.ac.uk/het/bentham/government.htm*. Although Bentham is sometimes credited with coining this phrase, a very similar expression was authored by the Irish philosopher Francis Hutcheson in 1725: "That action is best which procures the greatest happiness for the greatest numbers."

4. Bernoulli suggested that the utility of an amount of money diminished as the logarithm of the quantity, and logarithms do increase at a diminishing rate as a quantity gets larger. But there was no other basis for determining that the logarithmic approach actually quantified anybody's utility accurately.

5. Strictly speaking, utility theory can be used without game theory to make economic predictions, and it often is. But before game theory came along, the mathematical basis of utility was less than solid. In formulating game theory, von Neumann and Morgenstern developed a method to compute utility with mathematical rigor. Utility theory on its own can be used by individuals making solitary decisions, but when one person's choice depends on what others are choosing, game theory is then necessary to calculate the optimum decision.

6. See Ulrich Schwalbe and Paul Walker, "Zermelo and the Early History of Game Theory," *Games and Economic Behavior*, 34 (January 2001): 123–137.

7. The term "minimax" refers to the game theory principle that you should choose a strategy that minimizes the maximum loss you will suffer no matter what strategy your opponent plays and maximizes the minimum gain you can attain when choosing from each possible strategy.

8. In 1937, von Neumann published another influential paper, not specifically linked to game theory, that presented a new view on the nature of growth

and equilibrium in economic systems. That paper was another major element of von Neumann's contribution to economic science. See Norman Macrae, *John von Neumann,* Pantheon Books, New York, 1991, pp. 247–256.

9. In a footnote, he did mention possible parallels to economic behavior.

10. In the story, Moriarty appears in Victoria station just as Holmes and Watson's train departs for Dover, where a ferry will transport them to France. Watson believes they have successfully escaped from the villain, but Holmes points out that Moriarty will now do what Holmes himself would have done—engage a special train to speed him to Dover before the ferry departs. But anticipating this move by Moriarty, Holmes decides to get off the train in Canterbury and catch another train to Newhaven, site of another ferry to France. Sure enough, Moriarty hired a special train and went to Dover. But a game theorist would wonder why Moriarty would not have anticipated the fact that Holmes would have anticipated Moriarty's move, etc. See Leslie Klinger, ed., *The New Annotated Sherlock Holmes,* Vol. 1, W.W. Norton, New York, 2005, pp. 729–734.

11. Oskar Morgenstern, "The Collaboration between Oskar Morgenstern and John von Neumann on the Theory of Game," *Journal of Economic Literature,* 14 (September 1976), reprinted in John von Neumann and Oskar Morgenstern, *Theory of Games and Economic Behavior, Sixtieth-Anniversary Edition,* Princeton University Press, Princeton, N.J., 2004.

12. Robert J. Leonard, "From Parlor Games to Social Science: Von Neumann, Morgenstern, and the Creation of Game Theory, 1928–1944," *Journal of Economic Literature,* 33 (1995): 730–761.

13. John von Neumann and Oskar Morgenstern, *Theory of Games and Economic Behavior, Sixtieth-Anniversary Edition,* Princeton University Press, Princeton, N.J., 2004, p. 2.

14. Ibid., p. 4.

15. Ibid., p. 2.

16. Ibid., p. 6.

17. Samuel Bowles, telephone interview, September 11, 2003.

18. Von Neumann and Morgenstern, *Theory of Games,* p. 11.

19. Ibid., p. 11

20. Ibid., p. 12

21. Ibid., p. 14.

22. Ibid., *Theory of Games and Economic Behavior,* p. 13.

23. If you really want to get technical, you have to subtract the bus fare from the winnings (or add it to the cost) when calculating the payoffs for this game. But that makes it too complicated, so let's assume they live in a "free ride" zone.

24. Thus a mixed strategy is a "probability distribution" of pure strategies. The concept of probability distribution will become increasingly important in later chapters.

25. J.D. Williams, *The Compleat Strategyst: Being a Primer on the Theory of Games of Strategy,* McGraw-Hill, New York, 1954.

26. The actual math for calculating the optimal strategies for this game matrix is given in the Appendix.

27. In the original formulation of game theory, von Neumann insisted on treating games as if they were only one-shot affairs—no repetitions. In that case, a mixed strategy could not be implemented by choosing different strategies different percentages of the time. You could make only one choice. If your minimax solution was a mixed strategy, you had to use the random-choice device to choose which of the possible pure strategies you should play.

28. A similar version of this game is presented in a book on game theory by Morton Davis, which in turn was modified from a somewhat more complex version of "simplified" poker described by von Neumann and Morgenstern.

29. See Morton Davis, *Game Theory: A Nontechnical Introduction,* Dover, Mineola, N.Y., 1997 (1983), pp. 36–38.

30. Von Neumann and Morgenstern, *Theory of Games,* p. 43.

NASH'S EQUILIBRIUM

1. Roger Myerson, "Nash Equilibrium and the History of Economic Theory," 1999. Available online at *http://home.uchicago.edu/~rmyerson/research/jelnash.pdf.*

2. Paul Samuelson, "Heads I Win, Tails You Lose," in von Neumann and Morgenstern, *Theory of Games,* p. 675.

3. Leonid Hurwicz, "Review: The Theory of Economic Behavior," *American Economic Review,* 35 (December 1945). Reprinted in von Neumann and Morgenstern, *Theory of Games,* p. 664.

4. Ibid., p. 662.

5. Arthur H. Copeland, "Review," *Bulletin of the American Mathematical Society,* 51 (July 1945): 498–504. Reprinted in von Neumann and Morgenstern, *Theory of Games.*

6. Hurwicz, "Review," p. 647.

7. Herbert Simon, "Review," *American Journal of Sociology,* 50 (May 1945). Reprinted in von Neumann and Morgenstern, *Theory of Games,* p. 640.

8. In the film version of *A Beautiful Mind,* the math is garbled beyond any resemblance to what Nash actually did.

9. John Nash, "The Bargaining Problem," *Econometrica,* 18 (1950): 155–162. Reprinted in Harold Kuhn and Sylvia Nasar, eds., *The Essential John Nash,* Princeton University Press, Princeton, N.J., 2002, pp. 37–46.

10. John Nash, "Non-Cooperative Games," dissertation, May 1950. Reprinted in Kuhn and Nasar, *The Essential John Nash,* p. 78.

11. Ibid., p. 59.

12. Erica Klarreich, "The Mathematics of Strategy," PNAS Classics, *http://www.pnas.org/misc/classics5.shtml.*

13. Samuel Bowles, telephone interview, September 11, 2003.

14. Ibid.

15. John Nash, "Non-cooperative Games," *Annals of Mathematics,* 54 (1951). Reprinted in Kuhn and Nasar, *The Essential John Nash,* p. 85. I have corrected "collaboration of communication" as printed there to "collaboration or communication"—it is clearly a typo, differing from Nash's dissertation.

16. Kuhn, *The Essential John Nash,* p. 47.

17. As one reviewer of the manuscript for this book pointed out, it is not necessarily true that all economic systems converge to equilibrium, and that in some cases a chaotic physical system might be a better analogy than a chemical equilibrium system. The idea of equilibrium is nevertheless an important fundamental concept, and much of modern economics involves efforts to understand when it works and when it doesn't.

18. This observation (in a slightly different form) has been attributed to the physicist Murray Gell-Mann.

19. Quoted in William Poundstone, *Prisoner's Dilemma,* Anchor Books, New York, 1992, p. 124.

20. Mathematically, Tucker's game was the same as one invented earlier by Merrill Flood and Melvin Dresher. Tucker devised the Prisoner's Dilemma as a way of illustrating the payoff principles in Flood and Dresher's game. See Poundstone, *Prisoner's Dilemma,* pp. 103ff.

21. Charles Holt and Alvin Roth, "The Nash Equilibrium: A Perspective," *Proceedings of the National Academy of Sciences USA,* 101 (March 23, 2004): 4000.

22. Robert Kurzban and Daniel Houser, "Experiments Investigating Cooperative Types in Humans: A Complement to Evolutionary Theory and Simulations," *Proceedings of the National Academy of Sciences USA,* 102 (February 1, 2005): 1803–1807.

23. R. Duncan Luce and Howard Raiffa, *Games and Decisions,* John Wiley & Sons, New York, 1957, p. 10.

24. Ariel Rubenstein, Afterword, in von Neumann and Morgenstern, *Theory of Games,* p. 633.

25. Ibid., p. 634.

26. Ibid, p. 636.

27. Colin Camerer, *Behavioral Game Theory,* Princeton University Press, Princeton, N.J., 2003, p. 5.

28. Ibid., pp. 20–21.

29. The Royal Swedish Academy of Sciences, "Press Release: The Bank of Sweden Prize in Economic Sciences in Memory of Alfred Nobel 2005," October 10, 2005.

SMITH'S STRATEGIES

1. D.G.C. Harper, "Competitive Foraging in Mallards—'Ideal Free' Ducks," *Animal Behaviour,* 30 (1982): 575–584.

2. Of course, you could conclude that animals *are* in fact rational, or at least more rational than they are generally considered to be.

3. Martin Nowak, interview in Princeton, N.J., October 19, 1998.

4. Rosie Mestel, *The Los Angeles Times,* April 24, 2004, p. B21.

5. Maynard Smith's first paper on evolutionary game theory was written in collaboration with Price; it appeared in *Nature* in 1973. The story is told in John Maynard Smith, "Evolution and the Theory of Games," *American Scientist,* 64 (January–February 1976): 42. Price died in 1975.

6. John Maynard Smith, "Evolutionary Game Theory," *Physica D,* 22 (1986): 44.

7. Ibid.

8. The relationship between Nash equilibria and evolutionary stable strategies can get extremely complicated, and a full discussion would include considerations of the equations governing the reproductive rate of competing species (what is known as the "replicator dynamic"). A good place to explore these issues is Herbert Gintis, *Game Theory Evolving,* Princeton University Press, Princeton, N.J., 2000.

9. For the calculation of the Nash equilibrium giving this ratio, see the Appendix.

10. This equivalence of a mixed population—two-thirds doves and one-third hawks—with mixed behavior of the same birds holds only in the simple case of a two-strategy game. In more complicated games, the exact math depends on whether you're talking about mixtures of populations or mixtures of strategies.

11. Rufus Johnstone, "Eavesdropping and Animal Conflict," *Proceedings of the National Academy of Sciences USA,* 98 (July 31, 2001): 9177–9180.

12. John M. McNamara and Alasdair I. Houston, "If Animals Know Their Own Fighting Ability, the Evolutionarily Stable Level of Fighting is Reduced," *Journal of Theoretical Biology,* 232 (2005): 1–6.

13. Martin Nowak, interview in Princeton, October 19, 1998.

14. Ibid.

15. In all, 15 strategies participated in the round-robin tournament. Axelrod added a strategy that chose defect or cooperate at random.

16. Martin Nowak, lecture in Quincy, Mass., May 18, 2004.

17. A paper describing the results Nowak discussed in Quincy appeared the following year: Lorens A. Imhof, Drew Fudenberg, and Martin Nowak, "Evolutionary Cycles of Cooperation and Defection," *Proceedings of the National Academy of Sciences USA,* 102 (August 2, 2005): 19797–10800.

18. Herbert Gintis and Samuel Bowles, "Prosocial Emotions," Santa Fe Institute working paper 02-07-028, June 21, 2002.

FREUD'S DREAM

1. Von Neumann, actually, was very interested in the brain, and his last book was a series of lectures (that he never delivered) comparing the brain to a computer. But I found no hint that he saw an explicit connection between neuroscience and game theory.

2. P. Read Montague and Gregory Berns, "Neural Economics and the Biological Substrates of Valuation," *Neuron,* 36 (October 10, 2002): 265.

3. Colin Camerer, interview in Santa Monica, Calif., June 17, 2003.

4. Read Montague, interview in Houston, Tex., June 24, 2003.

5. The earliest MRI technologies were good for showing anatomical detail, but did not track changes in brain activity corresponding to behaviors. By the early 1990s, though, advances in MRI techniques led to fMRI—functional magnetic resonance imaging—which could record changes in activity over time in a functioning brain.

6. Read Montague, interview in Houston, June 24, 2003.

7. A Web page that tracks new words claimed that its first use was in the Spring 2002 issue of a publication called *The Flame.*

8. M.L. Platt and P.W. Glimcher, "Neural Correlates of Decision Variables in Parietal Cortex," *Nature,* 400 (1999): 233–238.

9. Read Montague, interview in Houston, June 24, 2003.

10. A.G. Sanfey et al., "The Neural Basis of Economic Decision-Making in the Ultimatum Game," *Science,* 300 (2003): 1756.

11. Read Montague, interview in Houston, Tex., June 24, 2003.

12. Paul Zak, interview in Claremont, Calif., August 4, 2003.

13. Aldo Rustichini, "Neuroeconomics: Present and Future," *Games and Economic Behavior,* 52 (2005): 203–204.

14. James Rilling et al., "A Neural Basis for Social Cooperation," *Neuron,* 35 (July 18, 2002): 395–405.

15. In this version of the game, Players A and B both get 10 "money units" and Player A chooses whether to give his 10 to Player B. If he does, the experimenter quadruples the amount to make 40, so Player B now has 50 (40 plus the original 10). Player B then chooses to give some amount back to A, or keep the whole 50. If Player A doesn't think B returned a fair amount, Player A is given the option to "punish" B by assessing "punishment points." Every punishment point subtracts one monetary unit from B's payoff, but it costs A one monetary unit for every two punishment points assessed. See Dominique J.-F. de Quervain et al., "The Neural Basis of Altruistic Punishment," *Science,* 305 (August 27, 2004): 1254–1258.

16. Colin Camerer, interview in Santa Monica, June 17, 2003.

17. Paul Zak, interview in Claremont, Calif., August 4, 2003.

SELDON'S SOLUTION

1. Werner Güth, Rolf Schmittberger, and Bernd Schwarze, "An Experimental Analysis of Ultimatum Bargaining," *Journal of Economic Behavior and Organization,* 3 (December 1982): 367–388.

2. Jörgen Weibull, "Testing Game Theory," p. 2. Available online at *http://swopec.hhs.se/hastef/papers/hastef0382.pdf.*

3. Ibid., p. 5.

4. Ibid., p. 17.

5. Steven Pinker, *The Blank Slate,* Viking, New York, 2002, p. 102.

6. Isaac Asimov, *Prelude to Foundation,* Bantam Books, New York, 1989, p. 10.

7. Ibid., pp. 11–12.

8. Ibid., p. 12.

9. Robert Boyd, interview in Los Angeles, Calif., April 14, 2004.

10. Some of the ultimatum game results were especially perplexing, in particular the first round of games played in Mongolia. Francisco J. Gil-White, of the University of Pennsylvania, was confused by the pattern of offers and rejections—until discovering that some players didn't believe they would actually receive real money. In another incident, he was puzzled by the rejection of a generous offer. It turned out the player thought Gil-White was an impoverished graduate student. By rejecting all offers, the player reasoned, he would ensure all the money was given back to Gil-White.

11. Joseph Henrich, telephone interview, May 13, 2004.

12. Colin Camerer, interview in Pasadena, Calif., March 12, 2004.

13. Ibid.

14. Robert Boyd, interview in Los Angeles, Calif., April 14, 2004.

15. Colin Camerer, interview in Pasadena, Calif., March 12, 2004.

16. David J. Buller, "Evolutionary Psychology: The Emperor's New Paradigm," *Trends in Cognitive Sciences,* 9 (June 2005): 277–283.

17. Not surprisingly, evolutionary psychologists have reacted negatively to Buller's criticisms, contending that he distorts the evidence he cites. You can find some of their counterarguments online at *http://www.psych.ucsb.edu/research/cep/buller.htm.*

18. Ira Black, remarks at the annual meeting of the Society for Neuroscience, Orlando, Florida, November 3, 2002. Black died in early 2006.

19. Steven Quartz and Terrence Sejnowski, *Liars, Lovers, and Heroes,* William Morrow, New York, 2002, pp. 41, 46.

20. E.J. Chesler, S.G. Wilson, W.R. Lariviere, S.L. Rodriguez-Zas, and J.S. Mogil, "Identification and Ranking of Genetic and Laboratory Environment Factors Influencing a Behavioral Trait, Thermal Nociception, via Computational Analysis of a Large Data Archive," *Neuroscience and Biobehavioral Reviews,* 26 (2002): 907.

21. Colin Camerer, interview in Pasadena, March 12, 2004.

QUETELET'S STATISTICS AND MAXWELL'S MOLECULES

1. Isaac Asimov, *Foundation and Empire,* Ballantine Books, New York, 1983 (1952), p. 1.

2. Ibid., p. 112.

3. Philip Ball, "The Physical Modelling of Society: A Historical Perspective," *Physica A,* 314 (2002): 1.

4. Ibid., p. 7.

5. Pierre Simon Laplace, *A Philosophical Essay on Probabilities,* Dover, New York, 1996 (1814), p. 4.

6. Gauss was not, however, the first to devise the curve that bears his name. The French mathematician Abraham de Moivre (1667–1754) initially developed the idea in the 1730s.

7. See Frank H. Hankins, "Adolphe Quetelet as Statistician," *Studies in History, Economics and Public Law,* 31 (1908): 33, 18.

8. Adolphe Quetelet, *Sur L'Homme et le developpement de ses facultes, ou essai de physique sociale.* The philosopher Auguste Comte also coined the term "social physics" about the same time, and had his own ideas about developing a science of society. See Roger Smith, *The Norton History of the Human Sciences,* Norton, New York, 1997, Chapter 12.

9. Adolphe Quetelet, Preface to *Treatise on Man* (1842 English edition), p. 7. Available online at *http://www.maps.jcu.edu.au/course/hist/stats/quet/quetpref.htm.*

10. Ibid., p. 9.

11. Ibid., p. 17.

12. Ibid., p. 14.

13. Ibid., p. 12.

14. Stephen G. Brush, "Introduction," in Stephen G. Brush, ed., *Kinetic Theory, Vol. I, The Nature of Gases and Heat,* Pergamon Press, Oxford, 1965, p. 8.

15. Henry Thomas Buckle, *History of Civilization in England,* quoted in Ball, "Physical Modelling of Society," p. 10.

16. Ibid., Chapter 3, "Method Employed by Metaphysicians," pp. 119–120. Available online at *http://www.perceptions.couk.com/buckle1.html.*

17. Ibid., p. 120. Note also that he allowed "experiments so delicate as to isolate the phenomena," but said that could never be done with a single mind,

which was always influenced by others, so that such isolation is really not possible.

18. Ibid., excerpt. Available online at *http://www.d.umn.edu/~revans/PPHandouts/buckle.htm.*

19. Ibid.

20. Quoted in P.M. Harman, *The Natural Philosophy of James Clerk Maxwell,* Cambridge University Press, Cambridge, 1998, p. 131.

21. James Clerk Maxwell, "Does the Progress of Physical Science Tend to Give Any Advantage to the Opinion of Necessity (or Determinism) over that of the Contingency of Events and the Freedom of the Will?" Reprinted in Lewis Campbell and William Garnett, *The Life of James Clerk Maxwell,* Macmillan and Co., London, 1882, p. 211.

22. Ignoring things like whether your opponent has a weak backhand.

BACON'S LINKS

1. www.imdb.com. There are additional actors in the database who cannot be linked to Bacon because they appeared either alone or with no other actors who had appeared in any other movies including actors connected to the mainstream acting community.

2. Similar network math was developed by Anatol Rapoport, who is better known, of course, as a game theorist.

3. Duncan Watts and Steven Strogatz, "Collective Dynamics of 'Small-World' Networks," *Nature,* 393 (June 4, 1998): 440–442.

4. Steven Strogatz, interview in Quincy, Mass., May 17, 2004.

5. These three examples were chosen because of the availability of full data on their connections; at that time, *C. elegans* was the only example of a nerve-cell network that had been completely mapped (with 302 nerve cells), the Internet Movie Data Base provided information for actor-movie links, and the power grid diagram was on public record.

6. Watts and Strogatz, "Collective Dynamics," p. 441.

7. In fact, here's a news bulletin: Oracle of Bacon hasn't updated its list yet, but as of this writing its database shows that Hopper has now surpassed Rod Steiger as the most connected actor, with an average of 2.711 steps to get to another actor versus Steiger's 2.712. Of course, these numbers continue to change as new movies are made.

8. Réka Albert and Albert-László Barabási, "Emergence of Scaling in Random Networks," *Science,* 286 (15 October 1999): 509.

9. Jennifer Chayes, interview in Redmond, Wash., January 7, 2003.

10. Thomas Pfeiffer and Stefan Schuster, "Game-Theoretical Approaches to Studying the Evolution of Biochemical Systems," *Trends in Biochemical Sciences,* 30 (January 2005): 20.

11. Ibid., pp. 23–24.

12. Suppose the drivers of two cars are caught on opposite sides of a snow-drift, both wanting to get through to go home but neither wild about shoveling snow. The "cooperator" would get out of the car and shovel through the snowdrift, while the "defector" would stay warm inside the car. If both defect, no snow gets shoveled and neither gets to go home, so they both lose. If they both shovel, they get to go home with half the work required of one shoveling alone. But if one shovels the whole thing, the other gets to go home for free. Game theory math shows that each driver's best move depends on the other's: If the other guy defects, your best move is to cooperate; if the other guy cooperates, your best move is to defect. This game is mathematically the same as the hawk-dove game in evolutionary game theory.

13. F.C. Santos and J.M. Pacheco, "Scale-Free Networks Provide a Unifying Framework for the Emergence of Cooperation," *Physical Review Letters,* 95 (August 26, 2005). A subsequent paper by Zhi-Xi Wu and colleagues at Lanzhou University in China questions whether it is the scale-free nature of the network that is really responsible for this difference, but that's an issue for further network/game theory research. See Zhi-Xi Wu et al., "Does the Scale-Free Topology Favor the Emergence of Cooperation?" *http://arxiv.org/abs/physics/0508220*, Version 2, September 1, 2005.

14. Holger Ebel and Stefan Bornholdt, "Evolutionary Games and the Emergence of Complex Networks," *http://arxiv.org/abs/cond-mat/0211666*, November 28, 2002.

ASIMOV'S VISION

1. In Sylvia Nasar's book *A Beautiful Mind,* she suggests that Asimov's Foundation might have been inspired by the Rand Corporation, where Nash worked on game theory in the early 1950s. But Asimov's novel *Foundation,* appearing in 1951, was in fact a compilation of short stories that had begun to appear before the Rand Corporation was created in 1948. The first Foundation story appeared in 1942.

2. Serge Galam, "Sociophysics: A Personal Testimony," *Physica A,* 336 (2004): 50.

3. The term "econophysics" was coined by Boston University physicist H. Eugene Stanley.

4. Serge Galam, "Contrarian Deterministic Effect: The 'Hung Elections Scenario,'" *http://arxiv.org/abs/cond-mat/0307404*, July 16, 2003.

5. James Clerk Maxwell, "Illustrations of the Dynamical Theory of Gases," *Philosophical Magazine* (1860). Reprinted in Brush, *Kinetic Theory,* p. 150.

6. An important point about statistical physics is that different microstates can correspond to indistinguishable macrostates. Various possible distributions of molecular speeds, for instance, can produce identical average velocities. That is one of the reasons why statistical mechanics is so successful. Even though it makes statistical predictions, in many cases the overwhelming number of possible microstates produce similar macrostates, so the prediction of that particular macrostate has a high probability of being accurate.

7. Katarzyna Sznajd-Weron, "Sznajd Model and Its Applications," *http://arxiv.org/abs/physics/0503239,* March 31, 2005.

8. Peter Dodds and Duncan Watts, "Unusual Behavior in a Generalized Model of Contagion," *Physical Review Letters,* 92 (May 28, 2004).

9. Steven Strogatz, interview in Quincy, Mass., May 17, 2004.

10. Colin Camerer, *Behavioral Game Theory,* Princeton University Press, Princeton, N.J., 2003, p. 465.

11. Damien Challet and Yi-Cheng Zhang, "Emergence of Cooperation and Organization in an Evolutionary Game," *Physica A,* 246 (1997): 407–428.

12. Jenna Bednar and Scott Page, "Can Game(s) Theory Explain Culture? The Emergence of Cultural Behavior Within Multiple Games," Santa Fe Institute Working Paper 04-12-039, December 20, 2004, p. 2.

13. Ibid.

14. Ibid.

15. Ibid., pp. 2–3.

16. Doyne Farmer, Eric Smith, and Martin Shubik, "Is Economics the Next Physical Science?" *Physics Today,* 58 (September 2005): 37.

MEYER'S PENNY

1. David Meyer, interview in La Jolla, Calif., August 6, 2003.

2. You can find more on this explanation for the quantum penny game in Chiu Fan Lee and Neil F. Johnson, "Let the Quantum Games Begin," *Physics World,* October 2002.

3. David A. Meyer, "Quantum Strategies," *Physical Review Letters,* 82 (February 1, 1999): 1052–1055.

4. David Meyer, interview in La Jolla, Calif., August 6, 2003.

5. Lan Zhou and Le-Man Kuang, "Proposal for Optically Realizing a Quantum Game," *Physics Letters A,* 315 (2003): 426–430.

6. This is a key point. You cannot use entanglement to send faster-than-light messages, because you need some other channel of communication to learn the measurement of the other particle.

7. Adrian P. Flitney and Derek Abbott, "Introduction to Quantum Game Theory," *http://arxiv.org/abs/quant-ph/0208069*, Version 2, November 19, 2002, p. 2.

8. Kay-Yut Chen, Tad Hogg, and Raymond Beausoleil, "A Practical Quantum Mechanism for the Public Goods Game," *http://arXiv.org/abs/quant-ph/0301013*, January 6, 2003, p. 1.

9. Azhar Iqbal, "Impact of Entanglement on the Game-Theoretical Concept of Evolutionary Stability," *http://arXiv.org/abs/quant-ph/0508152*, August 21, 2005.

PASCAL'S WAGER

1. E.T. Bell, *Men of Mathematics*, Simon & Schuster, New York, 1937, p. 73.

2. Blaise Pascal, *Pensées*. Trans. W.F. Trotter, Section III. Available online at *http://textfiles.com/etext/NONFICTION/pascal-pensees-569.txt*.

3. Laplace, a later pioneer of probability theory, did not find Pascal's argument very convincing. Mathematically it reduces to the suggestion that faith in a God that exists promises an infinite number of happy lives. However small the probability that God exists, multiplying it by infinity gives an infinite answer. Laplace asserts that the promise of infinite happiness is an exaggeration, literally "beyond all limits." "This exaggeration itself enfeebles the probability of their testimony to the point of rendering it infinitely small or zero," Laplace comments. Working out the math, he finds that multiplying the infinite happiness by the infinitely small probability cancels out the infinite happiness, "which destroys the argument of Pascal." See Laplace, *Philosophical Essay*, pp. 121–122.

4. David H. Wolpert, "Information Theory—The Bridge Connecting Bounded Rational Game Theory and Statistical Physics," *http://arxiv.org/abs/cond-mat/0402508*, February 19, 2004.

5. David Wolpert, interview in Quincy, Mass., May 18, 2004.

6. Wolpert, "Information Theory," pp. 1, 2.

7. E.T. Jaynes, "Information Theory and Statistical Mechanics," *Physical Review*, 106 (May 15, 1957): 620–630.

8. If you're flipping a coin, of course, the two possibilities (heads and tails) would appear to be equally probable (although you might want to examine the coin to make sure it wasn't weighted in some odd way). In that case, the equal probability assumption seems warranted. But it's not so obviously a good assumption in other cases. I remember a situation many years ago when a media furor was created over a shadow on Mars that looked a little bit like a face. Some scientists actually managed to get a paper published contending that the shadow's features were not random but actually appeared to have been constructed to look like a face! I argued against putting the picture on the front

page of the paper, insisting that the probability of its being a real face (or a representation of a face) was minuscule. But a deputy managing editor replied that either it was or it wasn't, so the odds were 50-50! I hoped he was kidding, but decided it would be wiser not to ask.

9. Jaynes, "Information Theory," pp. 620, 621.

10. This is, of course, the basis for teachers grading on a "curve," the bell-shaped curve or Gaussian distribution that represents equal probability of all the microstates.

11. David Wolpert, interview at NASA Ames Research Center, July 18, 2005.

12. Ibid.

13. Another decision-theory system was worked out by the statistician Abraham Wald, but the story of the similarities and differences between Wald's and Savage's approaches goes far beyond the scope of this discussion. If you're interested, you might want to consult a paper exploring some of these issues: Nicola Giocoli, "Savage vs. Wald: Was Bayesian Decision Theory the Only Available Alternative for Postwar Economics?" Available online at *http:// www.unipa.it/aispe/papers/Giocoli.doc.*

14. David Wolpert, interview at NASA Ames Research Center, July 18, 2005.

15. Strictly speaking, it's not the temperature of an individual, it's the temperature that the external scientist assigns to the individual, Wolpert points out. Just like in statistical physics, the temperature is a measure of what the external scientists infer concerning the molecules in a room (since a single molecule doesn't have any particular temperature—temperature is a property of the distribution of velocities of a set of molecules).

16. Wolpert, interview at NASA Ames, July 18, 2005.

EPILOGUE

1. Orson Scott Card, *Ender's Game,* TOR, New York, 1991, p. 125. Thanks to my niece Marguerite Shaffer for calling *Ender's Game* to my attention.

2. Ibid. p. 238.

3. Joshua Greene, interview in San Francisco, Calif., April 17, 2004.

Index

H

I

J

K

L

Skinner, B. F., 98
Small-world model, 149–153, 154,
   156, 157, 158
Smith, Adam, 9, 12–26, 31, 35, 78,
   106–107, 128, 219
Smith, Eric, 180–181
Smith, Roger, 11, 20–21
Snowdrift game, 162–163
"So long sucker" game, 61
Social cognitive neuroscience, 165
Social interactions. *See also* Social
   networks
   behavioral game theory and, 96–
      97, 108, 142, 174–175
   magnetism analogy, 169–173
   minority game, 175, 176–177
   modeling, 68–69
   molecular collision analogy, 153,
      166, 168, 173, 201, 210
   Nash equilibrium, 175
   opinion formation and
      transmission, 167–168, 169,
      171–173, 174
   pack/crowd behavior, 170, 171
Social networks
   acceptance of research on, 167
   clustering property, 154, 157
   contagion model, 173–175
   degrees of separation, 145–146
   evolutionary game theory and,
      159–160, 162–163
   growth of, 167–168, 224
   links between nodes, 148–149
   mathematical modeling, 159
   Nash equilibrium and, 166
   power laws and, 157
   small-world property, 151
   and statistical mechanics, 166
   terrorist, 167
Social physics, 244. *See also*
   Sociophysics
Social preferences, 111–112, 129

Social sciences, 3
   Buckle's philosophy, 137–138
   crime rates, 133–134
   and game theory, 30, 38, 50, 53,
      70, 119, 180
   Hobbes theory, 129
   long-term cooperative behavior,
      71
   metaphysical vs. scientific
      approach, 137–138
   physics and, 132–135, 142–143
   and statistics, 5, 129–132, 133–
      134, 138–139
Social validation model, 171–173
Sociobiology, 120, 223. *See also*
   Evolutionary psychology
Socionomics, 165
Sociophysics. *See also* Psychohistory
   computer simulations, 180
   cultural diversity and, 177–181
   and game theory, 175–177
   magnetism analogy, 169–173
   Nash equilibrium and, 60, 200
   networks and, 145, 163, 166
   and physics, 60
   probability theory and, 132–135
   Quetelet's average man, 133, 139
   resistance to, 166–169
   statistical mechanics, 142–143,
      165, 166, 168–169, 174,
      175, 199, 200, 210
   temperature of society/players,
      39–43, 165, 169, 173, 213,
      214, 249
Specialization, 25, 78, 108
Spite, 63, 111
Stability. *See* Nash equilibrium
Stag hunt game, 61
Stalemate, 172
Stanford University, 61
Star Trek: The Next Generation (TV),
   182–183, 188